T0159218

HOLT'S THEORY OF EVERYTHING

HOLT'S THEORY OF EVERYTHING

LET THERE BE LIGHT

JOHN R. HOLT

HOLT'S THEORY OF EVERYTHING
LET THERE BE LIGHT

iUniverse books may be ordered through booksellers or by contacting:

iUniverse
1663 Liberty Drive
Bloomington, IN 47403
www.iuniverse.com
1-800-Authors (1-800-288-4677)

ISBN: 978-1-5320-1453-6 (sc)
ISBN: 978-1-5320-1454-3 (e)

Library of Congress Control Number: 2017902062

Print information available on the last page.

iUniverse rev. date: 03/06/2018

CONTENTS

Part Four – Gravity

Part Five – Electromagnetism

Part Six-In Conclusion

FOREWORD

Have you ever wondered why light (energy) is seen as an electromagnetic wave when electromagnetic fields, immediately after the Big Bang occurred, where yet to exist even though energy (light) was in great abundance at this same time?

Have you ever wondered why Newton's theory of gravitation necessarily put the source of gravity for the sun and the planets orbiting around it at the center of these bodies?

Have you ever wondered why we can drop a lump of lead and a lump of aluminum from a high place and they will fall at the same speed? And to go even farther; in a vacuum, a ton of lead falls no faster than one tiny feather dropped alongside it. What is the relationship between matter and this mysterious power that can explain this?

Have you ever wondered what in nature, produces the effect of gravity?

Have you ever wondered how a photon, when emitted from its source, accelerates from zero to 186 thousand miles per second c without going any speed in between? And, how then, can this action really be classified as "acceleration"?

Have you ever wondered how a particle, traveling at 186,000 miles per second, can hit something and rather than cause extensive damage, just glance off and go in another direction while never slowing down?

Have you ever wondered why light travels so fast and what accelerated it to this speed to begin with and how it maintains this incredible speed?

Have you ever wondered how a photon in Young's double slit experiment knows whether the other slit is open or closed in determining if it will act like a particle or a wave?

Have you ever wondered why light can show itself as a wave or a particle depending upon which one you look for? And will light ever slow down or stop?

Have you ever wondered if the universe will eventually stop?

Have you ever wondered how the universe came to exist?

Have you ever wondered what Planck's constant really reflects in nature? And why it's so small?

Have you ever wondered why time slows for a body moving through space and why this same body contracts with this same movement? Or what would happen if a body, such as a space ship, actually reached the speed of light?

Have you ever wondered if all the forces could be reduced into one initial force?

Have you ever wondered if gravity will ever be unified with the three forces?

Have you ever wondered how the atoms making up everything in nature around us came to be?

Have you ever wondered if there is a true universal time we can use to reconcile all measurements, taken in relation to one event, by observers at different locations moving at differing speeds?

Have you ever wondered if there exists a basic particle in nature that makes up all other particles?

Have you ever wondered why the universe contains movement now when the second law of thermodynamics says it should have run down and stopped long ago? And just what mechanism or configuration allows it to defeat the second law?

Have you ever wondered just what causes inertia and does all matter possess this property? And why do all particles of mass have an unexplained property of gyre?

Have you ever wondered if a particle can have weight but no mass?

Have you ever wondered why the neutron is neutral?

Have you ever wondered if the neutrino has a value of mass?

Have you ever wondered why the electron, a negative particle, doesn't fly apart when all its inner parts, it is assumed, must be negative also?

Have you ever wondered why, when the atoms making us up are mostly space, when we clap our hands together, they don't go completely through each other?

Have you ever wondered how objects such as magnets can pull or push on each other with seemingly nothing in between them?

Have you ever wondered if the particle known as a gluon is in fact real?

Have you ever wondered whether wave theory or particle theory best explains the "sum of the parts "phenomenon relating to mass?

Have you ever wondered what happens to you when you die?

All the questions above and more will be addressed within the theory that I present. I hope I have aroused your curiosity enough to cause further interest in my model, for only by reading about it can you see how all the above phenomena are cleared up. I find it necessary to warn the reader that he or she will be wasting their time in trying to understand the theory I'm presenting within this book if they are less than acute in particle physics, theoretical physics and cosmology. I greatly doubt that this book will become a house hold favorite.

ACKNOWLEDGEMENTS

Within this model you will find works reflecting such "Classical" people as Anaxagoras, Democritus, Aristotle, Galileo, Newton, Huygens, Faraday, Maxwell, Planck and Lorentz. And, in the "Quanta/Classical" age I studied such as Einstein, Thomson, Rutherford, Bohr, Born, Dirac, Pauli, Lederman, Gell-mann, Heisenberg, Schrodinger, De Broglie, Glashow, "t Hooft, Schwinger, Yukawa, Weinberg, Fermi and many more than what I have room for here.

I will take time here to thank Indiana University for allowing me to post my partial works within Swain Hall. And I especially would like to thank Dan Beaker there for his patience in letting me bounce my ideas off him.

I do not include the science department faculty members within this work. I can only assume they were unprepared for a mental leap of the kind I presented to them. The same goes for others within the scientific community around the world. At this point in time and with reference to this particular field of study (2016 AD) I am most probably, the smartest human on this planet and I believe this work will eventually prove me out. I believe I have laid bare the path to solve this puzzle called the universe.

When I started this work in 1981, I did it out of a need to know just what the universe was and try to figure out what my role in it was. I had no idea that it would turn out as it did. But by use of empirical evidence and educated guesses, I feel that I have successfully figured out the basic design of an infinite and perpetual system, known as the universe. Be aware that my model discounts the idea of a single Big Bang in that such a configuration cannot fully explain how the universe began or how we got here. I do not make the claim that my model is absolutely correct in all it describes, but I believe it points the way. I hold it up as a lantern in the darkness and say, "Let there be light".

This work I am presenting covers my investigations between 1981 AD and year 2016 AD. Due to no one in the present scientific community being able or willing to give my work any credibility, I am aiming my theory towards the scientific community in the year 3000 AD. You might say that what I'm presenting could be called a scientific time capsule. I personally see it as the "Ultimate Occam's Razor" to include the old saying, "The simplest explanation is usually the best explanation".

With all that said, here's my theory in a nut shell: Existence consists of only energy which basically comes in two configurations, dormant energy (space) and kinetic energy (light).

Pressure is what converts dormant energy into kinetic energy (Big Bangs). Our Big Bang is only a local event producing plasmic kinetic energy of which some of the kinetic energy, produced by the Big Bang, produces quarks which, interacting with the fabric making up space itself, produce all the mass objects and forces around us in nature.

In the end though, entropy (space pressure) breaks all matter back down to just single quanta of radiation (light) which then can be reabsorbed by space to become part of the fabric of space once more. Within a dynamic configuration such as this, quarks are simply gravitationally bound up orbital particles of kinetic radiation which the present scientific community cannot conceive of yet.

I will now present one of the several incorrect assumptions the present scientific community holds fast to in hopes that this will arouse the reader's curiosity. With that said I offer up the "instantaneous acceleration of light" postulate for discussion. It is my opinion that acceptance of the instantaneous acceleration of light postulate can be classified as a major mistake made by science in that the belief in this postulate kept the true nature of mass and energy from coming out for so long. Observe the word acceleration within the postulate. Acceleration is the rate of change of velocity over time. But the postulate says "instantaneous" acceleration. The equation $d = rt$ fails here. Plugging the creation of light into this equation we get (distance greater than zero equals 186,000 mps times zero).

If we look at this logically we will see a conundrum. Time, in this circumstance, must equal zero in order to make the postulate feasible. This says that the universe must halt all motion within itself (time) while each particle of light changes speed from zero to 300,000 km/sec. After all, time at its base, is simply a measurement of actions during some specified period.

So how do I handle this conundrum? I simply assert that none of the particles of radiation ejected from the Big Bang ever moves thru space, or any other medium, at speed = zero until they are finally reabsorbed by the fabric of space itself. And this hypothesis can then be logically expanded to provide more evidence in reference to my opinion that what we see as mass is simply quarks which are produced by kinetic energy traveling in orbits about themselves within the quarks.

In conclusion, we can now come one step closer to understanding that all basic particles of mass (protons, neutrons, electrons, neutrinos, etc…) are simply quarks or combinations of quarks. I will show, throughout this work, how this all comes together in order to substantiate the reality surrounding us in nature. I would now like to invite you into my mind such as it is.

PART ONE – THE BEGINNING

REALITY

I feel I must convey to the reader, my concepts relating to what is real and what is illusion, in order to better let him/her understand the logic I used as I constructed this model. I do not challenge the wisdom or insight of such men as Kant, Nietzsche, Calvin, Hume, Hegel, Voltaire, or Bacon. I only admit that "something" exists. What that "something" is and how it works, is only speculation on my part. But an absolute right that humans possess, without question, is their right to speculate. Without speculation, the march toward knowledge would progress at a snail's pace if at all. So I speculate.

We as humans are basically constructed in a combination of what I call "Body supporting Mind". The human consciousness does not lie within the feet, chest or heart. These are simply some of the body parts, whose sole purpose is to support the brain. The brain contains the mind, which is who we really are. And the mind's function is to process the data it acquires from the body, concerning the nature of the environment around it, and then to formulate logical responses to the actions it senses within this environment. The mind gathers, analyzes and acts. The act can be either an immediate physical response or simply storing the analysis for future need. The point I am driving toward is, the mind is the divining rod that the human animal uses, to measure and assess the universe about him. Let us first analyze the tools the mind uses in its quest to understand.

We have been told since childhood that we have five senses; touch, taste, smell, hearing, and sight. These five senses are really just different manifestations of the one sense, which is touch. When we hear, vibrating air touches our eardrums. When we see, photons touch our eyes. When we taste, flavor touches our tongue. When we smell, aroma touches our olfactories. And when we touch, forces touch our body's nerve endings. The one common denominator here is "touch" and this is the only sensory-tool the mind uses, to probe the world around it. The brain (mind) gathers data through use of its sense of touch (body).

Humans subsist mainly by use of only one bodily organ and this is the brain. And this human organ has limitations such as the size of the brain, the speed of the brain, or the brain's ability to see the very small or very fast. An example of this is; the atoms that make up the matter in the world around us are roughly 99% space and 1% particles but we can't see this. And yet, light sources such as atoms cause photons of certain frequencies to come into contact with our eyes and the magnitude of these frequencies are electrochemically transmitted to our brain, where the brain then presents us with an illusion, called color. This says that color exists only in our minds. The mind uses color to artificially represent the form of an object it sees. If we could not see color but only the smearing actions of electrons in motion, we would never become aware of the Tiger lying in the weeds.

I included these thoughts, relating to what is seen and what is real, as an attempt to invoke upon the reader, a more open-minded attitude while reading the remainder of this work. In my analysis of reality, I found several striking pieces of knowledge and I will share some of them with you.

First off, we need to realize that what the mind allows us to see does not always reflect the true nature of the object, or action, in question. But the mind also has the ability to reach beyond the obvious in order to better let us understand the universe around us. This, we call constructive thinking. It must be understood that we as humans, with our limited intellectual capabilities, work off of a platform whose supports are girded with what I call "approximate knowledge". An example would be; take a candle, light it and then physically probe its flame with your bare finger. You find that the flame is hot but you do not know exactly how hot it is. That; is how you gain and work with "approximate knowledge".

Approximate knowledge lets some among us believe that as humans (objects of mass) we are in fact, a necessary component within the universe. That belief couldn't be farther from the truth. Energy is the sole physical entity the universe is comprised of. All matter beyond energy is just reactionary manifestations of kinetic energy (light) in motion thru dormant energy (space). Read my work and after some time you will understand.

I will now present a postulate whose base premise has, so far, been beyond human understanding. My postulate says, "We, as objects of mass, do not move thru space but move within space". In other words; we constantly "use" new space in our internal makeup as we go along. So in reality, all objects of mass consist of nothing more than kinetic energy (from the Big Bang) interacting within the physical space around it. Note: all the kinetic energy, which was ejected from our local Big Bang, is pressured "thru" space" but the objects of mass containing the kinetic energy making up the quarks within, move "within" that same space.

With all that said, do we now have a better understanding of reality? The dictionary describes reality as follows; "the totality of real things in the world, independent of people's knowledge or perception of them". This definition will suffice for what is to follow here. So let us now analyze reality. This analysis will be performed under a reference frame of the man made measuring tool called time.

The point within the first time frame we will choose to apply here is time $<10^{-34}$ seconds after the Big Bang occurred. How, under this circumstance, would we describe reality? Under the definition above, "the totality of real things in the world" at this point in time could best be described as a huge expansion of energy (see Gamow, Alpher and Herman). And what does this tell us?

It says, "the reality we know at present is basically a resultant product of primordial energy" which can be neither created nor destroyed and is still with us. Recall that under thermodynamics; all matter constantly radiates out into the cooler space surrounding it. In addition: the consequence of combining matter with antimatter is a resultant burst of gamma rays or energy, if you please. Any way you want to slice it, all present matter and photons still contain that same primordial energy. Note: at time $<10^{-34}$ seconds after the Big Bang occurred, mass had yet to evolve, so its conservation concerns were moot at that point. This, in itself, says that mass and energy in a chronological sense, are in fact not equivalent.

But more than that, this line of reasoning points to a certain conclusion implying; all mass matter, is simply different manifestations or configurations of "confined" kinetic energy in non-linear motion. If you have doubts about this then go and completely annihilate any elementary particle of your choosing and then watch as the subatomic binding properties (forces) within the particle are overcome and the kinetic energy "in motion" within the particle become free to escape in all directions in the form of gamma rays.

In conclusion; let me now show you an obvious result or idea the scientific community has consistently overlooked relating to reality before the Big Bang. If kinetic energy is the output after the Big Bang then logically some form of "potential" energy must have existed before the Big Bang. My best guess on what potential (static or unmoving) energy would look like in reality, would necessarily be the fabric making up space itself. This idea's time has come!

Let me cast about a little more with reference to reality. We, as objects of mass, are approx .001% kinetic energy and 99.999 reactive space. This line of thought led me to the realization that all bodies of mass, given time and under the influence of entropy, will eventually wear down, unmask and show their true identity which is kinetic energy. The energy making up the objects of mass and the universe as a whole are what is truly real for energy lasts forever and is the only necessary part of the universe. I would hope that by the year 3000 AD the scientific community can understand what I've proposed here.

In addition; we need to realize that many of the tools we use to analyze the universe around us were developed by humans and are not inherent in nature. Time, dimensions, mathematics, and laws, were all formulated by man in his quest to better understand and work with the universe around him. Man uses these, self-designed ideas, to help him with his investigation and the result is that he becomes somewhat more knowledgeable. On the other hand; the knowledge gained from these processes is only as good as the analysis process used. Quoting Bjorken, “Most theories, given time, will be disproved and are, at the least, incomplete”. This brings to mind the old adage; believe nothing you hear and only half of what you see. The wisdom of this becomes, “Question Everything!”

THE UNIVERSAL LAWS OF NATURE

I shall begin by stating that this model is based on three axioms:

Axiom # 1: IT IS IMPOSSIBLE TO OBTAIN SOMETHING FROM NOTHING.

Axiom # 2: IN ORDER FOR A THING TO EXIST (TAKE UP ROOM) IT MUST HAVE SOME PHYSICAL VALUE.

Axiom # 3: IF A SYSTEM HAS EXISTED FOREVER IN THE PAST, AND IS DOING WORK NOW, IT CAN BE CONSIDERED TO BE PERPETUAL IN ITS NATURE.

I need not proof these axioms for I believe they can be acknowledged as self-evident truths. It might appear to be a seemingly simple idea but I intend to show that important discoveries about the universe were arrived at, by applying the three axioms above to the physical world surrounding us.

The first thing I shall address is the difference between SOMETHING and NOTHING. Broken down these two words say, "SOME THING and NO THING". Or using a synonym we can write SOME SUBSTANCE and NO SUBSTANCE. We could also use matter for a synonym. To better let you understand these terms I will add that, all of these synonyms can be described by the definitions: SOME PHYSICAL THING for SOMETHING, and NO PHYSICAL THING for NOTHING. I can now make, with fair certainty that you will understand, the following proposition:

AXIOM # 1

NO PHYSICAL-THING CAN BE OBTAINED FROM
THAT, WHICH HAS NO PHYSICAL VALUE

I now present another proposition, which I also believe to be self-evident and as such can be considered an axiom. Axiom # 2 says:

AXIOM # 2

IN ORDER FOR A THING TO EXIST (TAKE UP ROOM) IT MUST HAVE SOME PHYSICAL VALUE

Using this axiom as a guide I find that the entire universe must be purely physical and no part of it can be considered to be nothing or void in its nature. In other words, the universe I speak of here includes everything that exists and even includes the space, which permeates throughin and throughout it. The scientific community, until recently, held the belief that "space" is basically void of all matter. If you really think about it, if space were truly void then it could not exist at all, let alone take up all the room that we know it does.

In addition, I find it rather interesting that the quantum theorists have had to resort to "borrowing" virtual particles from this "empty space" in order to balance their equations concerning black holes and high-energy collisions. The expressions dark energy and dark matter are also being thrown around due mainly to cosmological needs. String theorists are resorting to a similar plan of attack in their quest, to unify the forces. The Higg's field is their latest virtual device and is used in place of a physical space. History will show that both of these theories actually pointed to a physical space but the theorists showed an unwillingness to put their reputations on the line.

Axiom # 2 asserts that all things that exist within the universe must have some physical value in order to take up room. This axiom is all-inclusive and even applies to space itself. Let me reiterate: Anything that takes up room must have some physical value greater than zero in order to do so.

Using Axiom # 2 as a guide we find that, the universe is to the size that it cannot be considered to have an edge for if we were to consider what might be surrounding Everything (the universe), one could only come to one conclusion; Nothing! Since we have established before that only something can take up room, then we must conclude that anything that surrounds anything has to be something having a physical value. This also applies to anything within anything. Let me state the following: At no place either inside or outside the universe, can that, which cannot take up room, exist. The direct result of this axiom is that the universe contains infinitely smaller parts as well as being infinitely large as a whole.

Some hold the belief that both space and time were created simultaneously at some point in the past. They then postulate that space is bounded. But now the question needs be ask, "If space is seen as empty or nothing then what are we bounding it with?" Do we now have different kinds of nothing?

Isaac Newton, while analyzing the world around him, formulated what we call "the Laws of Motion". The third law of motion states, "For every action, there is an opposite and equal reaction". This law is also known as "cause and effect". If the universe is the effect, then we must assume that there exists a cause, and this cause must contain physical/

mechanical attributes. The main point to be shown here is; "That, which does not physically exist (Nothing), does not have either the means or opportunity to generate that which does exist (Something)".

And now I will present my third proposition as Axiom # 3.

AXIOM # 3

IF A "CLOSED" SYSTEM HAS EXISTED FOREVER IN THE PAST, AND IS DOING WORK NOW, IT MUST BE CONSIDERED TO BE OF A PERPETUAL NATURE

The Laws of Thermodynamics point to a conclusion that implies, "When all available energy within a closed system turns into unavailable energy the system will stop". As an analogy; when all the gas in a car runs through the engine and turns into exhaust, the car will stop and unless more gas is added, the car will remain stopped.

A question arises here as to whether the universe is open or closed. If the universe were open, energy within the system would be lost to the area outside the system. Within this model, there exists no outside area, since the system is infinite in size, which automatically precludes anything being around it.

Logically: The universe is a closed system with its closure being infinity (having no edge). You might also reason, since there is no outside area to lose energy to, the system would have to be considered closed. See Axiom # 2

The Thermodynamic Laws are the reason that you cannot get a patent on a Perpetual Motion Machine since, according to these laws, such a machine is impossible to make. I agree with this in principle for I don't believe that one can create a Perpetual Motion Machine, but I also believe the universe, we exist within, generates perpetual motion.

Let us consider the fact, using Axiom # 1, which says, the universe has existed forever in the past. Let us next consider the fact, the universe contains movement now. According to the Laws of Thermodynamics, the universe should eventually run down and stop and, since there is no other energy to add, it will stay stopped. I cannot accept the idea that something, which has already been running forever, might stop at some point in the future. The universe will continue to run (contain movement) in the future, for as long as it has in the past.

I submit that these axioms are inherent in nature itself and as such can be considered, Physical Laws of Nature.

AXIOM # 1

NO PHYSICAL-THING CAN BE OBTAINED FROM
THAT, WHICH HAS NO PHYSICAL VALUE
"You cannot get something from nothing"

AXIOM # 2

IN ORDER FOR A THING TO EXIST (TAKE UP ROOM)
IT MUST HAVE SOME PHYSICAL VALUE

AXIOM # 3

IF A SYSTEM HAS EXISTED FOREVER IN THE PAST, AND IS DOING WORK NOW, IT CAN BE CONSIDERED TO BE, OF A PERPETUAL NATURE

In combining the direct consequences of these three axioms, we end up with the following logical conclusions:

a. The universe, to include a physical space, has no true beginning (initial cause). It did not start, it always was. In addition; the motion it contains is perpetual else we could not be here now.
b. The universe is infinite in size and the terms, edge and center have no meaning here.
c. The universe is purely physical (solid); having no voids (areas that contain nothingness). Note here that under enough pressure, wave functions can be generated within a true solid.
d. Because the universe is purely physical, there cannot exist an area so large or small, that it does not hold something.
e. The universe cannot move external to itself, for it is already everywhere, or put another way; "There is no place that it isn't".
f. The Big Bang is only a local event, hence Hoyle's steady state theory, though somewhat modified, rises from the ashes like the Phoenix.
g. In nature (reality) there exists only one infinity: The pressurized universe. This, I believe, is the true basis for what is known as Space and Time. Space is the infinite physical universe and Time is the pressure driven interactions within the universe.
h. All interactions must be described in a mechanical framework and under motion in order to show actual happenings, for after all is said and done, the universe is simply, "A purely physical entity under pressure".

That concludes the first part of my model presentation and is separated from what is to follow simply by the fact, all that I have presented so far applies to the universe as a whole and can be proofed by the self-evident Axioms #1, #2 and #3. The rest of this model is also based on the three axioms just mentioned but in addition, it assumes that a Big Bang occurred within our local area in space and in fact is still occurring. Along with the above we also assume that space itself is a physical entity. As a side note, I must make it known that within this model the Wave/Particle Duality is not just an illusion but is as real, as you or I. Wave/ Particle Duality is the key to understanding all creation and I will show this.

With the exception of a particle-wave of energy (quanta) I use just one particle, which comes in various magnitudes of mass, to create all the other particles we have in nature around us. I at first described this unique particle as a Super Particle but upon researching high-energy physics through the works of such people as Lederman, Glashow, Pauli, Gell-Mann, and many others, I found my Super Particle was already on stage but under another name. Gell-Mann, with his reputation on the line, postulated that there might exist a particle (quark) that joins in combinations, to build the other particles around us in nature. At first, I doubted that the quark could do all that my Super Particle does, but upon further investigation, I discovered that Gell-Mann's quark was a physical equal to my Super Particle.

He found his quarks by force symmetry analysis, relating to high-energy physics, while I found mine by way of the same route, with the exception being, my high-energy physics involved the Big Bang. With this being said, I will discontinue the use of the term "Super Particle", within this model, and adopt Gell-Mann's "quark" in its stead.

I need to note that within this model, the word mass can be used synonymously with angular matter waves and also the quantum force and the terms, particles of space in motion, radiation, kinetic energy, photons and light, are all in fact identical in basic makeup and will be used synonymously within this work.

In addition, I will show that there was and still is, only one true force (pressure) active in the universe at large, with all other forces being just side effects (manifestations) of this one force. I can only hope that by "mechanically" explaining many of the phenomena that exist in science today, this model will be given serious consideration by the learned people within the scientific community.

GENESIS

In the beginning God created the heaven and the earth. And the earth was without form and void; and darkness was upon the face of the deep. And God applied his spirit upon the face of the deep while saying, let there be light, and there was light. And God saw the light, that it was good: And God then divided the light from the darkness. "And God called the light "kinetic energy" and the darkness he called "space". And the evening and the morning were the first day.

I did not quote the bible exactly as written but no really great differences did I inject into the bible's message. It seems that an oppressive darkness was everywhere before a force was applied and light was brought forth from the darkness after the force was applied. Sounds strangely familiar to another premise we call the Big Bang Theory doesn't it? So let us now explore more fully this line of reasoning.

The event known as the Big Bang, a local event, was caused by pressure acting upon the fabric making up space, as particles of light, submitting to entropy, transform into their "rest" state, which then causes the fabric making up space to rupture. We assume here of course, that the Big Bang actually occurred.

Within this model there is basically only one force, with which the universe operates and all other forces are manifestations of this ultimate force. This ultimate force is known as pressure and it permeates throughout the universe. Pressure not only showed itself when causing the Big Bang but also after the Big Bang. What existed right before the Big Bang was potential energy, the fabric of space, under pressure. What existed right after the Big Bang was the fabric of space and kinetic energy, the output from the Big Bang, under pressure. Pressure is a constant unrelenting force throughout the universe and this same pressure is also the main factor for the entropy the universe must undergo.

It may be hard to understand but what you see is what you get. In other words, what is making up the universe right now is actually the same as what was making up the universe both before and right after the Big Bang. Dormant energy (space) is simply light in its rest state. Kinetic energy (radiation) is bundles of light in motion (photons). And the only true force in all of existence (pressure) is produced by the situation where the universe contains more substance, cubic inch per cubic inch, than it can properly hold (See Perpetual Motion below).

PERPETUAL MOTION

At present, theoretical physicists and cosmologists are favoring two gravitational theories, black hole and reverse expansion, as their best candidates for building a return manifold between available energy and unavailable energy in order to explain how the universe works. Mathematically, reverse expansion would be the best choice of the two even though, as I will show, it is incorrect. On the other hand and under a configuration agreeable with the standard model, black holes must reach out with a particle called the graviton to capture matter and energy and pull them in.

But the main problem with both of these theories is the fact that neither can perform rejuvenation without their "open" system configuration losing available energy. Under the laws of thermodynamics, all open systems such as these, constantly radiate energy out and away from the system hence the systems will eventually run out of available energy. The model I'm about to present has no such problem with its return manifold system.

Once you have seen this model and even though you might not immediately understand parts of it, let me state, "It was designed in a completely mechanical/physical configuration". In my interpretation of the standard model for particles, I included only what was needed for creating an infinite perpetual universe.

With Occam's razor in mind I will state, "the universe as a whole is simply dormant energy (space), and kinetic energy (radiation) under pressure. This configuration allows the universe to be steady state and perpetual in its nature. Let me emphasize here that mass is nothing more than space reacting to kinetic energy orbiting within the quarks they make up. This will become clear once I show how that can be.

In addition; the mathematical value known as pi possibly hints at the hat trick the universe uses in order to attain its perpetual characteristic. It can be hypothesized that there exists .14159...... more substance in the universe cubic inch per cubic inch than it can properly hold. This puts a huge amount of pressure on space (dormant energy) which then responds with what we call Big Bangs.

My biggest problem in developing my model was figuring out how to defeat the second law of thermodynamics. Once I accomplished that it became clear; the universe exists in a condition such that the entropy it must undergo is the very cause for its rejuvenation.

Quoting from pg. 9 in Steven Hawking's book, A BRIEF HISTORY OF TIME, "A theory is a good theory if it satisfies two requirements: it must accurately describe a large class of observations on the basis of a model that contains only a few arbitrary elements, and it must make definite predictions about the results of future observations". I intend to show that my model will more than satisfy these requirements.

One side note I shall insert here is that a man by the name of Alpher, along with Gamow and Herman, proposing radiation as the initial Big Bang output, was one of the keys that helped me formulate this model.

Mathematically; and given enough time, a steady state configuration is the only possible descriptor for an infinite and perpetual universe. Within this model the explanation for the universe's continual internal movement is; the condition exists whereby the universe actually contains more physical substance, cubic inch per cubic inch, than the room it takes up can hold, and space "as a physical entity" is included also. At first glance this statement seems ludicrous, but let me add one more parameter to this condition.

I state again that the universe actually contains more physical substance than it has room to hold, but now I want to subject some of this substance to movement and behold, Lorentz's contraction comes into play and causes the substance to shrink in size the faster it moves. What is shown here is that the universe can contain itself, only so long as some of what it contains is put into motion. I assert again; the universe exists in a configuration such that the entropy it must under-go is the very cause for its rejuvenation.

Again; there exists in nature only one original force from which all other forces are derived and that force is the one we call pressure. This model shows that the generator of this one force is also the recipient of its reaction, which is motion. Put simply, "the universe, under extreme pressure, perpetuates a degree of movement within itself". The mean rate, of this movement, just happens to be the speed of light.

I find that I cannot construct a completely comprehensive model without discussion of what shall be described as an "Initial Condition" for the universe. I want to note that my dealing with this initial condition will necessarily be "hypothetical" as, an infinite universe with no beginning or end and containing perpetual motion, cannot truly have an initial condition.

Humans as logical creatures, become perplexed when faced with the idea that there might exist something having no beginning or initial cause. The universe, under this model, did not start and has always existed in a state of containing some equivalent of movement. This idea can be stated thusly; Pressure begets Forces which begets Actions which begets Motions.

Allow me to wax philosophical on this point. If I was challenged to create a Perpetual Motion Machine and had all the necessary ingredients, this is how I would construct it. First I would take a bowl, made of pure matter that contains no movement, and pack it full of more pure matter that contains no movement, to the point to where I could absolutely not, fit anymore matter in the bowl. Then I would cover the bowl, with a cover made of this same matter. I will note here, I still have some pure matter left over, but this is all the matter left in existence except for the matter making up the bowl and all that's in it.

And now I would make the bowl, the cover, and all that the bowl contains, infinite in size. What I would have at this point is a completely static (unmoving) and infinite universe. This condition (configuration) of the universe might have been possible except for the fact that we know that the universe presently contains movement. I will hypothesize here that if the universe were truly static (no forces interacting) then pi would be equal to 3. In addition, under this circumstance entropy would equal zero.

I will use the Second Law of Thermodynamics as a corollary for stating, "If the universe was ever of the configuration just described (inert) it would still be in this same condition, or state, now". This law implies, "If a closed system is stopped and there is no available energy to apply to it, the system will remain in this stopped condition". With that aside let us now try and start our machine.

I take the cover back off my bowl and insert a super sized mixer and start stirring. When I have everything mixing with a high degree of motion I put the cover back on and step back to watch. The second law of thermodynamics (entropy) immediately sets in to stop my machine. I then watch as all the actions within my machine start slowing down and eventually all motion stops.

So you can see that I need to do something more to this machine. The problem here is, the only thing left in existence, besides the bowl and all that's in it, is the extra amount of pure matter that wouldn't fit in the bowl in the first place. And my machine is completely full with no open areas anywhere within it. But I have an idea I want to try.

I take the cover back off my bowl and reinsert the super sized mixer and start stirring again. The rate that I choose in my stirring is 186k miles/sec (the speed of light). I continue to stir until a goodly portion of everything is moving at a high speed (speed of light). I then notice a strange thing happens. All this matter I'm stirring at the speed of light starts shrinking in size (Lorentz's Contraction). So I put in more matter (remember the matter I had left over) and let it shrink, and then more and more until at last, I reach the point to where I have no more matter left to add. It turns out that the extra matter I added equals .14156.......of all the matter (cubic in per cubic inch) within the bowl. And now, this bowl, which is infinite in size, contains everything in existence to include itself.

Now, I quickly put the cover back on and stand back and watch. Sure enough, the Second Law of Thermodynamics starts applying entropy to my bowl and everything in motion within it. Entropy is simply, space applying pressure to all the kinetic light particles making up matter. I realize at this point that all things must submit to entropy and expect to see all the movement I put in the bowl, slow down and stop. But I find that a strange thing is happening within my bowl. Remember that my bowl, its cover, and everything within the bowl make up an object, which is infinite in size. Just for the heck of it, let us call this object the "universe". Now, as entropy attacks everything within this universe, to break it down, I notice that when the extra matter I added to the universe starts to break down and stop, it necessarily starts to re-expand (Lorentz's expansion). I realize that entropy is trying to return all this matter back to the configuration it was in before I stirred it. So what exactly is the end result here?

The Second Law of Thermodynamics (entropy) is itself supplying the pressure, which the universe uses to defeat it. As entropy acts on the matter within this universe, the matter re-expands, but the situation is finally reached where there is simply not enough room to hold all this re-expansion. Remember when I put the extra matter in while this universe was being stirred. And remember that my universe (bowl, cover and everything within the bowl) was infinite in size. Well, this means that I created the situation where the universe actually holds more matter, cubic inch per cubic inch, than it has room for, except when some of this matter is under motion.

Entropy cannot be evaded but entropy, acting on the matter making up this bowl and everything within it (the universe), is what brings about the extreme pressure needed to generate a rupture in the matter itself (Big Bang). The pressure has become so great that the universe tries to "explode" but the problem here is, something infinite in size (having no outside), has no place to "explode" into. So finally it turns to its only alternative, which is to "explode within". And here it all goes again, and again, and again, without end. That's how I would construct a Perpetual Motion Machine. **The beauty of this approach is the fact that it presents a possible explanation for how we and the rest of existence can be here now**.

Is anything possible within an infinite universe? The answer is definitely, yes! If we mathematically look at possibilities as a set, then we necessarily must include impossibilities within this set, in order to make the set whole. With this philosophical idea being stated, I will confess that I have constructed my Perpetual Motion Machine, hypothetically, with the end result being proofed by an impossibility (See Axioms #1, #2, and #3).

This Perpetual Motion Machine I created was all done in my mind but one just like it does exist. This machine I speak of is the universe itself for it has no beginning, it always was and it always will be.

The reason that the universe we live in contains motion at present is because its physical configuration defeats the Second Law of Thermodynamics. I designed this one by using the force of pressure. I tried many other approaches but all configurations of causes led me right back to pressure. It was only when I put together an equation, whose factors were dormant energy (space), kinetic energy (light) and force (pressure), could I give a possible explanation for the universe as it is today. The guidelines I observed were the Axioms #1, #2, and #3, and the foundation was built upon the Big Bang and Standard Model. In addition, Planck's Constant, the Lorentz/FitzGerald Contraction, the Kinetic Law of Gases, the Second Law of Thermodynamics, deBroglie's Pilot waves and Newton's Third Law of Motion, play an immensely important part in this model. In addition; within this work the letter c when presented alone is meant to represent the speed of light and the characters (, ^,), when combined as (^) is meant to represent the point in our local area of space where the Big Bag occurred.

I hope the fact is realized by anyone reading this model that the present state of the universe is given a "possible" description within this model, up to this point. I would reiterate; the Perpetual Motion Machine (the universe), I speak of above, is so special that it didn't have to be designed or manufactured; it's been here forever and running for just as long.

THE MASS CREATION PERIOD

The result of the Big Bang (space rupturing) was a huge plasmatic ejection of energy in the form of radiation (light). This is the reason I show mass as a result of particles (quarks), being made with particles of light, and not the other way, as it is assumed at present. In addition; because all forces, masses and photons have a base energy value tied to them (Planck's constant), which is directly tied to light, I had to make light the base particle (factor) in relation to them all.

Now we will move slightly forward in time and we see that the Big Bang, using the force of space rupturing, accelerated the particles making up the fabric of space, which under motion, contract and become radiation, to very high speeds, which "initially" exceeded the speed that we now believe particles of light attain, 186,000 miles per second c. The density of the space in the area where the Big Bang occurred and the magnitude of the pressure exerted to generate this Big Bang are the two factors, which governed the energy output of the Big Bang, and determined the initial speed with which the radiation (light) was ejected. In this situation, the speed of light was "not" independent of its source.

The immense heat (magnitude of radiation/volume) from the huge amount of energy being released from this local area of space during the Big Bang caused the surrounding fabric of space to be greatly expanded. This would allow particles of light to move at a speed in excess of c, at least until they meet up with denser normal space and allows us to say that gravity is now on stage. To reiterate this point; if pressure and radiation is all that exists immediately after the Big Bang then that's basically all we have to work with. Hence, the kinetic energy is the direct generator or cause for all gravitational fields (expanded space).

As the leading wave of ejected light particles started meeting up with normal denser space, they start decelerating back to speed c. Meanwhile the light particles behind this leading wave, while still running hot, started overrunning this leading wave, causing a piling effect. This was the beginning of the mass-particle creation period. I will show; there exists only one mass-particle, which by its various configurations, makes up all other particles. This basic mass particle is known as a quark.

Let us now assume the Big Bang occurred. Then let us assume Gamow, Alpher and Herman were correct in their premise that the initial ($< 10^{-34}$ seconds) output from the Big Bang event was primarily a huge outpouring of plasmic kinetic energy. Quoting Encarta 2003; "they theorized the universe was very hot at the time of the Big Bang". And we know of course that thermodynamically, the word hot points directly to a presence of radiation. And this is the point where we "hypothetically" pause or stop everything for analysis. So, using just these two assumptions above let us attempt to reveal the order under which the "assumed" known forces evolved.

Adhering to the assumptions above, we can safely presume that an action of some great enormity has occurred. But what we need to try and analyze here is the impending reaction, for that is where the origin of the forces resides at this time. And if we give some small amount of consideration to what we have here we must realize that this line of thought begs the question; just how can there "not" be forces present at this particular point in time, to allow for the generation of the future forces. The answer is obvious if we just take time and really think about it. If we assume that the Big Bang can be described as a great release of energy then we can now assume with little doubt, this great release of energy would necessarily be accompanied by a release of pressure of some sort.

At present, most within the scientific community see the Big Bang as a reaction to some, as of yet, unknown cause. And though we might not know the exact cause we can be fairly certain of its result which can, in all probability, be described as a gargantuan release of energy. But we now ask, how does all this super expanding energy produce a force?

It has been found that light possesses a property known as momentum and it has been shown that a single particle of light can exert a small amount of pressure on any surface it comes into contact with. Hence, there must exist a category of force called radiation pressure. And what conclusion can we draw from this?

With the exception of the causal pressure and the reactive plasmic radiation, the first force, present at $< 10^{-34}$ seconds after the reactive event known as the Big Bang, would had to have been a result of radiation pressure. And if we analyze each of the four assumed forces in order to pick the most favorable force, which can possibly fit with what is happening here, it becomes more than obvious that gravity is the clear choice. And this allows me to conclude that gravity and radiation are definitely connected.

There can be little doubt that the mechanical actions inherent within Dr. Einstein's General Theory along with the huge radiation pressure force being applied should by now be starting to affect the physical configuration of space itself in order to generate what will be known as a gravitational field (the lair where gravity lurks). And this says that the second "assumed" force to come onto the stage is gravity. But note that I, like Dr. Einstein, do not recognize gravity as a force. In addition, I see no possible configuration of forces at this early point which could validate the graviton theory. Again, if all that exists at this very early stage is radiation and space itself, then gravity would more than likely be the result of radiation interacting with the fabric of space itself.

To continue; radiation which has already come on stage, is believed to be the force transmitting boson relating to electromagnetism. But I will show later in this work, electrons (which have yet to come on stage) in motion, is the main generator for the creation of electromagnetic fields. Now I will advance time to what we will call the Big Bang nucleosynthesis period.

According to present theory; the electromagnetic, weak and, strong forces are transmitted by bosons. The third force to come on stage would logically be the quark generated gluon which is thought to have the ability to defy the conservation of energy law. I will refrain at this point from telling the reader what I think of that hypothesis. The force transmitting boson for electromagnetism is believed to be the photon which I addressed above so the electromagnetic force, accordingly, comes on stage fourth. This puts the weak force in fifth place.

In addition; I do believe there exists a sixth force which is life itself. All bodies containing life are constantly in action and action is a prime factor in describing forces, hence, if it walks like a duck and talks like a duck then it must be a duck (force).With all this said it still must be understood that under this model all forces are basically just reactive expressions of the one real force which is pressure itself. You ask, how can it be that simple and I reply, it just is.

Immediately after the Big Bang and during the release of the pressure that space was under, the force of pressure was still being applied to the radiation that had been forced into motion by the Big Bang. As time evolved this same space pressure became instrumental in both maintaining and regulating the speed, at which particles of light travel. The pressure being applied to the top, bottom, sides, and rear of the particles of light, force the particles of light along while the frontal pressure regulates their speed to equal c.

But this action has an opposite and equal reaction (Newton's 3rd Law of Motion). Imagine contracting a sponge to the point that it can no longer absorb water, and then injecting a drop of water into it at some pre-defined speed and direction. Since the sponge cannot absorb this drop of water, yet is applying pressure to it from all sides, the drop is forced along at near the same speed and direction it was injected with. This sponge analogy is not intended to represent actual events, but is only used to give the reader an idea (approximate knowledge) of what is actually occurring.

Space's reaction, to the motion of particles of light moving through it, is the generation of a kinetic wave that projects just ahead of the particles of light. This "pressure-sustained" kinetic wave, when traveling in orbits within quarks, is the generator of mass. A particle of light in motion through space creates an effect (kinetic wave) we define as energy (action). This is the reason for the phenomenon called the Particle/Wave Duality and it is real. But more than that, this shows us the ultimate equation (F=Ec). E or Energy within this portion of the universe, moves at the speed c and F or Force, is simply the fabric of space reacting to this energy in motion. The energy of a single particle of light equals Plank's constant and it manifests itself as the basic unit of all forces. Put simply, kinetic energy is the "action" and force is the "equal and opposite reaction" (Newton's Third Law of Motion). Without energy there would exist no matter or matter interactions. With the exception of space pressure, kinetic energy is the base action behind all forces. Without energy nothing can exist, not even space itself. See Light on pg. 51.

The first two properties within particles of mass we shall address are gyre and inertia. The particles of light moving in their orbits within the quarks give the quarks a value of gyre (internal gyroscopic motion). This action of gyre gives quarks a stabilizing effect (inertia) the same as a toy top achieves when spun. Note that my model explains something here (gyre/inertia) that science has been aware of for years but could not explain. In addition; the particles of light moving in their orbits within the quarks angularly drive the space surrounding the quarks which gives the quarks an angular property of force I call matter waves. And these matter waves become what we call mass. In addition, these matter waves take away the need for the magical particle known as gluons. With the exception of anti-quarks, the matter waves keep the quarks from coming together and annihilating each other and the gravitational fields generated by the kinetic energy orbiting within the quarks keep the quarks from wandering off on their own. For those of you in the far future, be aware that mass is presently seen as a particle property.

As the particles of light travel in their orbits within the quarks they make up, they physically drive the space from the area within the quarks. Space pressure, because of this kinetic action within it becomes much lower here than outside the quark and this generates an "inertial guiding path of least resistance" within and around the quarks. This is the direct mechanical cause for approx. 99.999% of the gravitational field surrounding an atom. And the strength of this gravitational field at this quantum level is extremely stronger than it is outside the atom. When I say that space pressure is much lower within and around the quark, I mean that space is being kinetically expanded so that the magnitude of kinetic action within and around the quark is much greater, than it is out away from the quark. Quantum gravity's magnitude here is great enough, to collectively capture particles of light and force them into extremely small orbits, in order to form what we call quarks.

When more than one quark joins together to comprise larger particles the mass of each quark is added together to give us total mass. There is however, a possible exception to this for sometimes empirical investigations (high energy collisions) seem to show a slight loss of total mass when compared to individual masses. Mass (spatial wave interactions) produces this exception. Within this model mass is an energy generated angular force (effect) within space itself. Keep in mind that all the above speaks to mechanical processes relating to the generation of quark gyre/inertia and quark mass. This approach keeps everything real.

A heavy quark, having more particles of light in its makeup, has more weight and mass than the lighter quarks. When a quark and an anti-quark annihilate, the product of this annihilation is gamma rays (light). In the modern day version of quantum mechanics, when a particle meets its antiparticle they annihilate each other and the result is a burst of gamma rays. This detail, I will use to support my claim which says that all mass objects contain kinetic energy (light). In addition, there is a thermodynamic law, which concludes that all matter radiates. Since, within this model, quarks are presented as the basic particle of mass, from which all other mass particles are constructed, it follows that all the radiation, with two exceptions, must come directly from the quarks. It does not matter whether a quark radiates or undergoes full or partial decay, for the output is the same, namely photons. The exceptions I speak of above relates to the atom's ability to absorb and radiate light and of course all the light from the Big Bang (background radiation) which was never captured in order to create mass particles.

In addition, interpreting mass as a unit of weight is incorrect. The unit that applies to mass must be one that reflects the physical properties of an angular force located in the fabric of space immediately surrounding the quarks and must also include, as a base factor, Plank's constant. A force can be defined as a physical influence that can change the speed or direction of motion of an object it comes into contact with. In presenting an analogy of this process, imagine that you walk at 3 miles per hour straight towards a merry-go-round, which is spinning at 20 miles per hour (outer edge), and you step up on the merry-go-round. You will find that this action causes you to ultimately end up out in the yard somewhere away from the merry-go-round. What actually took place is the result of a linear force being applied to an angular force. This is why, when photons, having no mass, hit us at speed c, we are not tore asunder. Mass, which this model delineates as an angular force or matter wave surrounding quarks, resists nearly all forces, linear or angular, brought against it. The exception is when an anti-quark interacts.

Particles of light by use of their associated kinetic waves physically stretch or expand the space surrounding them and this action generates what is known as a gravitational field. Gravity is simply an "inertial guiding path of least resistance" being offered by the expanded space surrounding bodies of mass. Each particle (Planck) of radiation generates a quantum gravitational field, which by itself is insignificant in the classical sense, but when combined in large numbers within a very tiny area, such as they are within quarks, a "local" quantum gravitational field of considerable strength is created. This is what I call "Quantum Gravity".

As the particles of light and their associated kinetic waves move through their orbits within the quarks, space is driven away from the quarks in an angular direction (matter waves). This action causes a low space pressure condition within the quarks. This low space pressure is also known as an inertial guiding path of least resistance. Because space is kinetically expanded at this quark level, this implies that this is also where the gravitational field, per its size, will be the strongest. I will state," The Inverse Squared Law for gravitational fields does not hold for the area within the source for gravitation (quarks)".

Gravity rules the planets, stars and galaxies but electromagnetism can overpower gravity on our classical level. However, at the atomic level gravity once again rules, but at the sub-atomic (quark) level it is once again overpowered, for a short distance, by the quark "matter waves" (angularly-biased quark matter waves or what this model sees as mass force). These "angular" matter waves, if you remember, are generated by the light traveling thru their orbits within the quarks which drives the fabric of space surrounding the quarks. This is the exact repelling mass force that keep quarks from colliding with each other as they interact. But this repelling force between the quarks is limited in its range (approx.10^{-20} cm). Farther than that and gravity once again takes over which keeps the quarks in a 3 quark package (proton) configuration.

Note: this allowed my model to do away with the fictional particle known as the gluon. In addition, my model now shows the exact mechanical cause for what is called the "strong force". The "strong force" is simply "quantum gravity". When an atom comes into contact with a gravitational field the kinetic particles of light in motion within the quarks, electrons and neutrinos making up the atom, find it easier to work in a direction towards the source of the gravitational field rather than away from it. This is the real description for gravity. At present all known forces are seen and defined as active in nature but within my model gravity is an inert or passive force, if you will, that resides in the physical configuration of space itself within the gravitational fields.

The fact that protons contain only 3 quarks is because the heavier quark with the most mass can capture and orbitally retain only two lesser quarks. All three quark's matter waves constantly repel each other as they do their dancing around each other. Hence, the strength of the proton's 3 quark matter waves (masses) keeps all other quarks from physically joining in. Do you remember what I said before about quarks being the basic mass particle from which all other mass particles are made? With this in mind let us return to the time just after the Big Bang occurred.

If we assume that Gamow, Alpher and Herman were correct in their Big Bang proposal making kinetic energy the one and only output from the Big Bang then we must also assume, under this model, that a huge gravitational field (expanded space) is being generated as all this energy is ejected outward. Remember what was said before; that gravity is a result of space being stretched by light as it passes thru it which creates an inertial guiding path of least resistance within space itself (gravitational field). Now, as all this energy is expelled from space in the Big Bang we see that its initial speed is greater than what we have come to call the speed of light c. This is due to the amount of pressure applied to space in order to make it transform (Big Bang) and also due to the huge amount of expansion that space is undergoing. Hence the leading wave of energy is running fast as it spreads out thru the space around it in all directions.

You can say that this particular action is tachyonic in nature. But, eventually this leading wave of energy starts to meet up with more normal less stretched space. This slows the speed of the leading wave but the result of this is; the trailing waves of energy which are still running at a speed in excess of c start over-running the leading wave. This generates a piling-up action between energy moving at speed c and energy moving at a speed in excess of c. To repeat, this is the beginning of quark (mass) creation. As the two separate energies tangle up with one another the gravitational fields they are both generating take hold and force them into orbits around each other. We now have quarks. Note that quarks have no nucleus within. I will state, "All truly elementary particles have no nucleus". Within this model only quarks fit this narrative.

At first, when we are dealing with this huge amount of energies interacting, many heavy quarks are created, but as time progresses not so many heavy quarks are created but many less massive quarks are. This is the time where quark matter waves and gravity rules and as I will show, protons start appearing. At that point I will show how my model produces these protons and also electrons, neutrinos and neutrons along with their anti-particles.

As this "quantum mechanical" quark creation period progresses and piled up energies becomes more scarce, the creation of the heaviest quarks dies off and the creation of lighter quarks begin. One thing of importance I must address here is the physical/mechanical configuration of all these quarks. As all these energies pile up on each other they pile up in different configurations. You might say that some quarks are right handed and some are left handed. You can also say that right handed quarks absolutely do not like left handed quarks (quarks vs anti-quarks). The mechanical configuration relating to this is the case where the direction that light moves within a quark is the same for all the particles of light within that quark. Then we get a quark that is identical except that the direction the light moves is opposite from the other quark. If these two quarks meet, their matter waves, rather than repulse, actually join. Quantum gravity is at play here. But this is not good for either quark. All symmetry and balance of internal forces are lost, hence annihilation happens. When they meet each other and this occurred frequently right after the Big Bang, it was total annihilation with only gamma rays (light) left over. For some reason, I don't pretend to understand, only one, either the left handed or right handed quarks, ended up being the dominate quarks left in our part of universe. Under this theory there could possibly be anti-galaxies out there.

Middle ways thru the quark creation period a lighter quark was created we know as the electron. This quark (electron) has much less light in orbits within it than does the heavier quarks. Scientists have long been puzzled over how a negative particle, which they "wrongly assume" is made of internal negative parts, holds together. I show within this model that the supposed negative, positive and neutral charge properties of elementary particles have nothing to do with holding the particles together. The present practice of giving charge values of positive, negative and neutral is inappropriate. The inner "kinetic" physical configurations of the quarks within the particles determine how the particles will react when subjected to an electromagnetic field, not the charges they are supposed to possess. My model has solved for another problem here. Again; this explains how an electron, being considered all negative, to include its "unknown" constituent parts within, can be stable and not fall apart. When an electron is brought thru a magnetic field its biased directional matter waves interact with the electromagnetic field and consequently the electron path bends toward the positive. Depending upon their momentum and how much change of direction they experience a value of mass for the electron (quark) can be calculated. When an elementary particle moves thru an electromagnetic field, a situation develops where two different kinetic configurations of space itself interact with one another. More on this later within this work See Gravity on page 143.

At the end of the quark creation period one last quark was created we know as the neutrino. Some in the scientific community conjecture that this super light neutrino (quark) might be absent of mass. It does come close but because it's a quark, it has particles of light in orbits within it, hence it has some value of mass. This neutrino (quark) is very special for its quantity throughout the universe determines how many heavy nuclei can exist. Let me explain.

Quarks have particles of light in orbits within them all going in the same direction. But there is an exception and it is the neutrino. This quark contains the least amount of light a quark can have and still exist. In fact, light is so scarce within this quark you can have its particles of light within, going in two directions at the same time. I configured this by using an extension of Pauli's exclusion principle.

If physicists can ever subject a neutrino to a magnetic field all they will see is a squiggle as the neutrino goes through the field. This is because the dual mechanical matter waves (left handed and right handed) generated by the neutrino and then encountering the magnetic field will force the neutrino first left and then right as it goes through (squiggle). Note here that I convey to you the fact that the neutrino has no anti-particle. At present within the scientific community, no one knows (mechanically) why, except to say, "it's just neutral and that's all". As you can see my model mechanically goes all the way in physically explaining why the neutrino can have no anti-particle. In addition; the neutrino is not the only mass particle that squiggles when it goes thru a magnetic field, the "neutron" squiggles also and to a greater degree.

Within this quantum mechanical model the neutron consists of 5 quarks (3 quarks for the proton, 1 quark for the electron, and 1 quark for the neutrino). The quantum mechanics relating to these mass particles within the neutron goes like this: At or near the end of the quark generating period gravity was hard at work trying to gather all particles of mass together. Most attempts failed but sometimes, when symmetry and forces balanced, higher order particles appeared and prospered. The proton, electron and neutrino were the most successful particles that came out of the Big Bang. What is spectacular about these 3 particles is the fact of how they could be fit together. First you must find an electron that has used its gravity to trap a neutrino which then takes up an orbit around the electron. The neutrino, you might say is both positive and negative at the same time (neutral matter waves). This allows the neutrino to almost get past the electron's matter waves and physically join with the electron, but it fails. However; the effort does bring it close enough to be trapped by the electron's gravitational field. This, in effect, gives this electron/neutrino (nulectron) a physically neutral configuration when compared to all the other particles. Then along comes a proton and its larger surrounding gravitational field, hunting, looking, feeling, searching, until it meets up with this nulectron. Because the electron has a neutrino in orbit about it this, in effect, makes the nulectron neutral. Now, same as the situation above with the neutrino and electron, the nulectron is able to get close enough to the proton so that the protons gravitational field captures it. A new particle is born and its name is, "neutron".

As an aside: particles of light traveling in orbits within quarks, angularly drive the space surrounding the quark, as they move around their orbits. This driving of space by the light particles sets up an angular wave in space (quantum force matter wave or just kinetic energy wave). This means that quarks radiate matter waves in a spherical 360^0. The matter waves sweep out in an expanding configuration. Using a proton as an example; these waves that are angularly radiating out from the three quarks within the proton meet up with each other's waves and repulsion between the three quarks is the result. This is the mutual properties belonging to quarks that keep them from coming together and annihilating each other. On the other hand, if a quark having angular matter waves opposite to another quark it meets (anti-quark) they will complement each other and quantum gravity will cause them to come together, tangle up and annihilate each other. I believe galaxy interactions show some of these same traits though on a much larger scale. I've looked up galactic collisions on the computer in order to analyze my hypothesis and it seems to have some merit. If their spiral arms are in motion in an opposing configuration (same rotational configuration) when the galaxies come into contact they fight. If their spiral arms complement each other (opposite rotational configuration) they gravitationally join or combine much more easily.

The gravitational field within and surrounding a nucleus keep the nucleus from flying apart as the quantum forces (angular-biased quark matter waves) interact. This same gravitational field within and around the earth keeps it from flying apart also. The equation is simply, the more particles of light an object contains, the greater is the expansion of the space within and around it, and hence, the larger is its associated gravitational field (inertial guiding path of least resistance). The fact that gravity is an "inertial guiding path of least resistance" is the direct reason this "so-called force" refuses to unify with the other forces. This is because gravity is not an active force at all, and can actually be considered, as a lack of force (or an inert/passive force for you mathematicians). Gravitational fields (result of expanded space) simply presents a guiding path of least resistance to all the particle/waves of energy making up all of creation.

In truth, the gravitational field, because of its physical configuration, cannot act (exert force) upon anything at all. It must be acted upon. See Gravity on page 143. It is well known that force equations relating to the electric and magnetic fields do not hold when applied to the gravitational field. This is due to a remarkable property belonging to gravity such that, bodies that are put in motion, through only the influence of a gravitational field, receive an acceleration that has nothing to do with either the total weight or mass of the object in motion. A simple adjustment to the force equations now being used can rectify this unification problem. It must be noted, when an object is put into motion by a gravitational field, there is no "accelerative force" by the field itself, to factor, because no "accelerative force" by the field is being applied. In fact; the object being accelerated is causing its own acceleration. "This is a consequence of an inertial guiding path of least resistance" interacting with objects of mass.

The scientific community presently sees gravity as almost a non-player at the subatomic level but we must look beyond the obvious here for this is the place where the gravitational fields are omnipotent. Consider the mass we might calculate for the top quarks, and convert this to its associated gravitational potential while factoring in the size of the top quark. This in itself says that the gravitational field of this top quark must be very significant indeed "below the Fermi level". A close analogy of the interaction between mass and a path of least resistance would be the aerodynamics associated with an aircraft and the air around its wings as it flies. The wings are designed to create a vacuum (low air pressure) above them as the plane flies through the air and this creates a path of least resistance for the wings of the aircraft. This process is understood by some, as resulting in the plane being pulled upward as it flies, when in fact, it is simply easier for the wings to rise than it is for them too fall. There are less atoms of air (pressure) directly above the wings than directly below, so the plane rises. And the path of least resistance is the cause for this. I need to point out that the association between mass and gravity is not quite, as it is believed at present, but for this case it will suffice.

Particles of light with their associated kinetic wave, while traveling through their orbits within quarks, generate angular (left handed and right handed) matter waves around the quarks, which are the source of the quantum force (matter waves). These angular matter waves are what keep the quarks from coming together and annihilating each other. These angular matter waves are also the direct cause for what some are calling mass. Again; within this model, mass is basically an inherent property of matter and is generated as a result of particles of light traveling thru their orbits within the quarks which, by the way, make up all other matter. In addition; the electron and its associated matter wave, when moving at high speeds within its orbit around the nucleus, generates a strong kinetic wave (DeBroglie) that physically drives the space ahead of the electron. Note: when a voltage is applied to a conductor such as a copper wire the electrons within the wire are set into motion along the length of the wire. The actions of the electrons result in a reaction by the space surrounding the wire which goes by the name electromagnetic field. More on this later.

Physicists at present are still trying to understand the consequences connected to Pauli's exclusion principle. This model proposes a simple configuration of electron interaction in order to shine light upon this problem. Basically; as space is driven ahead of the electrons (deBroglie) as they travel thru the space ahead of them, surrounding space then moves in behind the electron to rectify this difference in space density. This is the action responsible for creating standing waves around the nucleus of atoms. And this is why only one electron is able to take up residence within a hydrogen atom. The electron and its frontal kinetic wave drives space ahead of it which then generates a resultant standing wave in the space behind. The resultant standing wave eventually dissipates with time and distance but until then its physical presence can be felt and it will physically rebuff any advances made by foreign electrons that might happen by. In close to the nucleus, the standing wave (quantum force) will repel any attempt made by a foreign electron to move in and bind with the nucleus.

I need now to go somewhat farther here with this particular arrangement. If we assume the electron possesses the physical characteristics of a sphere, which the standard model sometimes proposes, then we must determine how the electron's trailing standing wave repels all foreign electrons approaching in all directions. This forces us to propose a standing wave design where the generated standing wave's force is sustained well beyond just several orbits around the nucleus. The beauty of this approach is that it allows for a telling mathematical picture to be created with only a slight extension to Pauli's exclusion principle. The picture looks like this: Shortly after the Big Bang, particles of light pile up on each other and their individual gravitational fields bind them together to form quarks. Next, particular three quark combinations gravitationally combine at a very small distances from each other, which creates protons. The individual matter waves of the quarks, generated as the particles of light orbit within the quarks, keep them from physically joining and annihilating each other.

As an aside; under this line of reasoning it is obvious that neutrons have yet to appear on stage. As the mass creation period of the Big Bang continues a lighter intermediate quark appears. This quark goes by the name "electron". The electrons are captured by the proton's gravitational field but cannot physically join due to the repulsion generated by their individual matter waves. But as this occurs, a significant action is taking place. The electron cannot escape the gravitational field of the proton and it also cannot overcome the proton's quantum forces (matter waves). Hence it ends up in an orbit close to but not within the nucleus. This arrangement creates what is called hydrogen. For quite some time after the Big Bang hydrogen was the only atom produced. This was because the protons, which were plentiful, and because of their matter waves, could not join in any stable configuration. This situation, I calculate, continued until the neutrinos came on stage. So now the neutrinos were gravitationally captured within orbits around the electrons to produce a neutral particle called the nulectron (Once again as before; the quantum forces interacting between the electrons and neutrinos keep them separated while their gravitational fields keep them together in a configuration comprised of neutrinos orbiting electrons). This sets up a situation where this neutral particle (nulectron) can join with the protons and this produces neutrons. Neutrons, being neutral, now allow protons to join in a configuration whereby protons and neutrons can join together to create the nuclei belonging to the heavier elements in nature around us. Note; I have no doubt that Novas produce the heaviest elements present in nature today.

The scientific community at present, sees the protons as a source for the theorized positive charge and the electrons as a source for the theorized negative charge, and have been puzzled by the equality of these charges. It would make sense that the proton, being approx. 1836 times more massive than the electron, would have a larger net charge than the electron. But this is not the case. And what does this say? It says, charge is, by its nature, a phenomenon, directly tied to the physical/mechanical characteristics acting within matter. Charge is assumed to come in only two forms, positive and negative with electrons determining whether an atom is seen as one or the other. But what about neutrons? Thru high energy physics it has been shown that neutrons contain both positive and negative particles of charge. In addition, it has been shown in cloud chambers that neutrons squiggle as they move thru electromagnetic fields. Under this model this can be mechanically explained as neutrons are basically made of a proton (three heavier quarks), electron (one intermediate quark) and neutrino (one super light quark). The neutrino gives the neutron a neutral charge but the neutrino, which orbits the electron, which orbits the proton does not always 'fully' show itself to the electromagnetic field. Hence, the electron is able to partially show itself at times to the electromagnetic field. This why a neutron squiggles in a cloud chamber as it moves thru an electromagnetic field.

In addition; when an electron is put in motion it spawns a kinetic wave that rides just ahead of the electron (DeBroglie's pilot wave). This wave then, as it pushes thru the space around it, generates an electromagnetic field within that same space. I've never really accepted this idea of charge mainly because no true "mechanical explanation" has ever been given for it. Within this model I present here, the positive, negative and neutral characteristics of atomic particles reside, not within the charges but within the particles themselves. As I proposed previously; the physical/mechanical configuration within the quarks involved, determine how the particles will travel after being subjected to an electromagnetic field. What I'm pointing out here is; charge is just a theorized property within the atomic world that, as far as I'm concerned, really doesn't do much of anything without an electromagnetic field being present.

Early in the 20th century (1919) Albert Einstein proposed that electrically charged particles might possibly be held together by gravity. Consider that the nucleus contains nearly all the mass within an atom but can only neutralize the negative charges belonging to the electrons surrounding it. This fact is telling when related to sub-atomic investigation. At present it is believed that the differing charges hold the atoms together. The problem with this approach becomes apparent if the atoms are subjected to strong electromagnetic fields. If local electromagnetism holds the atoms together then an introduction of a strong foreign electromagnetic force should very easily tear the atoms apart. But we know this is not the case. Chemists presently use a configuration involving the outer electrons when explaining atomic interactions responsible for binding atoms together to form molecules. This approach necessarily sets up a circumstance within the molecules making up elements, where the charges of the atoms involved are forced to go from neutral to negative quite often, which would then leave the atoms as sitting ducks for any foreign forces that might be present within the local area around the atoms. Under Einstein's proposal noted above, no such binding problem occurs. This says, "at the lower subatomic level the electromagnetic force must eventually bow to gravity".

A kinetic wave is located just ahead of the electron and fits the description of DeBroglie's pilot waves. The electron and its frontal kinetic wave affects (drive) the fabric of space it encounters as it orbits the nucleus. This is how the electrons might add some small value of both mass and gravity to the atom. Science has yet to come up with a reasonable explanation that might explain how elementary particles constantly maintain their spin. In addition, these same particles have been found to possess a very physically powerful property of stableness we call gyre. Spin and gyre are two different properties belonging to the elementary particles. Within this model the explanation for this is: the particles of light orbiting within the quarks produce the property of gyre. And the quark, as a whole, revolves which produces the property of spin. To put it simply you might say; the electron's matter waves (quantum force) drive space out and away from the nucleus. Any radiation that enters the area affected by the strength of these mater waves will be forced to act accordingly and this "accordingly", will depend upon the internal makeup of the atom.

In addition, we can say that the electrons, orbiting a nucleus and driving the space surrounding the nucleus, pull a vacuum of sorts on the nucleus. The nucleus exists in a state of low space pressure (high gravity). And this high gravity characteristic (inertial guiding path of least resistance) allows electrons into orbits around the nucleus where the electron's kinetic standing waves determine the shells and number of electrons in each of these shells. This is basically how Pauli's exclusion principle works. The only exception to this is the 1rst shell where the nuclear force (nucleus matter waves), which acts for a short distance only, interacts with gravity to determine the 1rst shell's distance from the nucleus. Basically, it is the gravitational field generated by the nucleus which determines the size of the atom, and the strength of this field is determined by the number of nucleons in the nucleus. As an aside; de Broglie postulated that," the motion of a particle is governed by the wave propagation properties of certain "pilot waves" which are associated with the particle. These pilot waves were experimentally confirmed and led to the idea of "standing waves" that electrons generate as they orbit the nucleus of an atom. This then led Heisenberg to his famous "Uncertainty Principle". The "Schrödinger Equation" also reflects these pilot waves and is an integral part of what we now call "quantum physics". If one analyzes the Fourier Method that was used to combine the effects of all the singular wave and group wave characteristics of an electron, it soon becomes obvious that the standing waves cannot be a rear-generated wave, but must be the reactive result of a forward "projected" wave.

Particles of light, traveling in their orbits within quarks, produce a gyroscopic effect (gyre) we know as inertia. The unit of measurement called mass is the measure of an object's matter waves (quantum force). All forms of matter, with the exception of the particles making up the fabric of space, individual particles of light, and groups of light particles moving together in a linear direction (photons), have a value of mass. In photons, the particles of light making them up herd together, and move in linear directions. There is no gyroscopic property associated with them. This is why photons have no mass. I need to make it clear that photons are a man made idea that he applies to the samples of light he investigates. If photons had mass they would have a huge adverse effect upon anything they came into contact with. The effect a gravitational field has upon matter is connected to the individual actions of the particles of light, within the matter. Photons, having no matter wave characteristics such as objects of mass possess, travel thru electromagnetic fields in straight lines. This little fact has been in plain sight for all to see for at least a century.

As an aside; all things have weight, which is directly related to the number of particles of light they contain and their location within the gravitational field they are in. If a body is not within a gravitational field its weight would necessarily be zero. The gravitational field doesn't act upon objects per se, but only offers an inertial guiding path of least resistance to the individual (Planck) particles of light within the objects. It can be said here that the individual particles of light within bodies of matter act upon the gravitational field, not the other way around. This is why, heavy and light objects in a vacuum, fall to earth at the same speed (See Gravity pg. 143).

Dr. Einstein, in 1915, refined his calculations to show the degree that light would be deviated from its path of travel, when in the vicinity of a massive object such as the sun. He believed that this was a result of the sun's gravitational field acting upon the mass of the light, going by it. The problem with this belief has to do with light having a value of mass and also his assumption that the gravitational fields act upon mass. Within this model, a particle of light, while traveling in a straight line through space, has no value of mass. But it does have weight, and this weight can be calculated by comparing the strength of the gravitational field around the sun, to the degree of path deviation observed as the light passes it. The frequency of the photon under measurement will not affect the accuracy of the result, as the gravitational field does not physically recognize the photon as a whole but only the individual (Planck) particles of light that become entangled within it. The equation would necessarily have to match the deviation observed, as light passes by the Sun (Dr. Einstein's calculation).

Under Dr. Einstein's relativity, all objects having a value of mass and when under motion, experience what is believed to be time dilation. It must be recognized here that the term known as time is, in reality, a man made idea used to measure motions within the universe around us. This time dilation effect really has not so much to do with time, but more to do with the changing orbital rate of kinetic energy within the quarks making up the object, when under motion. If we were to find an object, truly stationary in space (with reference to the area in space where the Big Bang occurred (^)), the particles of light moving in orbits within the quarks making up this object, would do so at the speed c. When this object is in motion through space, these particles of light, which are only allowed to move at the rate of c must share their speed between their orbits within the quarks and the speed the object is moving through space. Hence, it takes longer for the particles of light to complete their orbits within the quarks, which then diminishes the strength and frequency of the matter waves generated by these quarks. Note: light moving in an orbit and also in motion thru space (O=>) shares its motion.

An item of interest that falls out of this model has to do with Einstein's proposal that an object's mass increases, the faster it moves through space. My model shows that mass (angular force) actually decreases, the faster an object moves, but at the same time, this object contracts

(Lorentz's Transformation) and becomes much more dense in relation to its size. This is why two protons, when colliding at near the speed of light, seem to be more massive. Using this information and accelerating a proton to near the speed of light and then directing it at an electron does little damage to the proton but has dire consequences for the electron. Relativity, as presently understood, says that the electron and proton do not know who is moving and at what speed during this event, but when we reverse this process we see right away that this does not hold up. I will go into more detail on this later within this model. The main point here is that any object of mass, when put into motion, will experience a loss of mass. If an object of mass were to reach the speed of light all the kinetic energy within the quarks making up that object could no longer maintain an orbit about each other but would now move off in straight lines thru space. This action does away with the quarks (mass) within the object and in effect leaves only photons behind. This says simply that mass is simply an effect of corralled kinetic energy (light).

Some believe that mass can be converted into energy and that energy can also be converted into mass. I find this idea to be a false hypothesis. This model holds that, mass is actually a spatial effect generated by kinetic energy traveling in orbits within quarks. To put it simply; when mass disappears only light is left. I will be more than disappointed if you students in the 31rst century have yet to discover this fact. On the other hand, the only way energy can be converted into particles of mass is during an event such as the Big Bang, possibly novas and high energy physics (colliders). High-energy physics has produced events where mass production, seems to have been accomplished, but I would rather think that partial quark decay, could account for this. This I surmised, due to the fact that, these "new" mass particles soon decayed. A badly wounded quark (slow decay rate) would show characteristics similar to what might be thought of, as mass production. When mass particles such as protons are slammed together at high velocities, their quarks can experience immediate decay. The quarks can also break apart and the resulting partial quarks can exist for a short time, but because of internal force symmetry imbalance, they decay into photons rather quickly. The particles produced from the Big Bang, which is the ultimate high-energy situation, produces stable quarks of the kind we see making up the world around us.

We can now return to the point in time just after the Big Bang and we find radiation being ejected at a super velocity. We also find that this same pressure that caused the Big Bang is still here, and applying itself to these particles of light. This is the cause for the speed with which light travels. But this action has an opposite and equal reaction. At the same time space is accelerating the particles of light thru itself it is also restricting the motion of the particle of light. The result of this duel action between space and light is that a constant speed c by the particles of light is attained. In addition, as these particles of light are squeezed through space, this same space reacts by forming a kinetic wave just ahead of the particles of light. This is the mechanical explanation for light having both particle and wave characteristics. The particle/wave duality of light is not an unexplainable phenomenon but is in fact, the true nature of light. The particle of space put in motion at the Big Bang is the action. Space, generating a kinetic wave (radiation force wave) just ahead of the particle of space in motion through it, is the reaction (3rd Law again). A German chemist Ostwald argued that energy is the "unique" real material in the world, and matter is not a carrier, but a manifestation of it. I agree, with the stipulation that this energy originally comes from space itself. We can now return again, to the point where I was explaining how mass particles are created.

As I said before, the trailing edge of the radiation over runs the leading edge of the radiation ejected from the Big Bang. As these particles of light are forced into non-linear motion around each other, they physically drive the fabric of space ahead of them. This is the mechanism that generates matter waves around the quarks. I call this the quantum force, and is not to be confused with what is now called the strong nuclear force. The strong nuclear force is "assumed" to be an attractive force where the quantum force within this model is repulsive.

Within this model all forces, excluding gravity, are basically the result of driven space. On our classical level, quantum gravity is seen as a tiny force, but at the sub-atomic level quantum gravity is actually a huge inertial guiding path of least resistance. Einstein published a paper around 1918 postulating that atomic particles are held together by gravitational forces. 't Hooft, in speaking of quantum gravity, pointed out that gravity, is a force at the quantum level and be it large or small, there is no theory for it at this level. I am now fixing that problem.

The quantum gravity keeping quarks together in a threesome configuration (proton) has not been discovered or recognized by particle physicists, and they assign the binding properties of attraction between quarks to an "imaginary" force carrying particle, called the gluon. The gluon is seen as a particle with the peculiar ability of being able to increase its force strength, over distance. I have never seen or heard of the Law of Physics, allowing this phenomenon to occur. In Newton's 3rd Law, the reaction must be opposite and equal to the action, and in this case would mean that the reaction generates added force as it gets farther away from the original action. This goes even beyond the idea of action at a distance for it now presents "increased" action at a distance. My model had to exclude the gluon for obvious reasons.

In actuality, the quarks are held together by quantum gravity and held apart by the angular matter waves (quantum force) emanating from the quarks themselves. The matter wave strength decreases over distance very quickly but the gravitational strength extends well out beyond the proton. When the three quarks combine (proton) and reach stability, they exist in a condition where the matter wave strength (quantum force) and the gravitational strength (quantum gravity) cancel each other which prevents annihilation. But when the quarks are shocked into separation such as during high-energy events, their quantum force and quantum gravity is overcome and this in effect, causes the demise of the proton. This action is the underlying reason for the confusion that led to the theorized gluon and its strange power of increased force over distance.

Up to this point we have examined how the mass generating particles (quarks) were formed. Basically, we have particles of light making up the two classical forms of energy responsible, for the configuration of nature around us. These two forms (also particles if you will) are the quarks and photons. In the book, "The Second Creation" on page 395 Salam ask the question, "Why does nature have to use both, quarks and leptons?" Both the quark and photon come in quantized values, depending upon the number of particles of light in their makeup. So we must remember that light itself is the base value or factor, in all that we investigate concerning nature around us. Quarks contain particles of light in non-linear orbital motion while photons contain light in linear motion.

We must remember that the amount of radiation ejected from the Big Bang is finite and this says, that as the quark (mass) creation period advanced in time, less and less particles of "piled up" light, became available for quark creation. This is why there are so many varieties of quarks. The first quarks created were what we call the top quarks. These are the heaviest or most massive quarks, as they contain the maximum number of particles of light allowed in the orbits within the quarks. These heavier quarks rule or control the other quarks they will join with, in creating the nucleons (protons and neutrons). All nuclei have heavier (more massive) quarks within each of its protons and neutrons and these heavier quarks rule the domain of the nucleus simply because, the strength of their quantum forces and quantum gravity, are the greatest of all the quarks. It is possible that heavier quarks formed particles, by joining together in the beginning of the mass particle generation period but these heavier particles created would be short lived, as the intermediate quarks, generated later, were drawn in to join with the heavier quarks at a much closer distance, which would disunite the attractive force between the heavier quarks. Force symmetry plays a major role at this time. Right after the Big Bang, if interacting kinetic energy forces don't balance the particle fails to exist for any notable length of time. If the interacting forces balance out then we will have longer lived entities, namely the heavier quarks.

Now, as time moves on and particles of "piled up" light become scarcer, the intermediate quarks are created. These will join with the heavier quarks and their configurations will be determined by the strength of the angular matter waves they possess. The heavier quarks, due to their having more particles of light in orbits within them, will have stronger angular matter waves (quantum force) and also a stronger gravitational field (quantum gravity).

The heavier quarks have a stronger gravitational field than the intermediates do. This is what makes the heaviest quarks, at this subatomic level, the Big Dogs. It holds the intermediate quarks to it with its gravitational strength (quantum gravity), while also keeping them at a distance with its angular matter waves (quantum force). The intermediates also possess this same quantum force and quantum gravity. The quarks, with the exception of three quark bundles called protons, normally do not physically interact with each other except under high energy events which leads to annihilation. This occurs when two quarks, having opposite going angular matter waves, are brought together by their gravitational fields (quantum gravity). An anti-quark is created, by simply turning a quark 180^0 vertically. This is why so much annihilation took place immediately after the Big Bang. Some like to think of anti-particles as negative energy, mass, or matter. As you can see from what I have just shown, this is a misnomer. Annihilation is experienced when numerous particles of light making up quarks come into physical contact with one another.

At this point we will examine the property quarks possess called spin. In this model, spin by a quark as a whole can vary, depending upon the nature of the quark involved. The spins of the three individual quarks within protons can change, depending upon the interactions between the quarks. The spin of electrons stay mainly stable throughout. Neutrino spins can be erratic since the kinetic energy orbiting within them goes both ways. Note: in a three-quark configuration containing a heavy quark and two intermediate quarks (proton), all the quarks have angular matter waves of basically, the same orientation. This sets up the condition where the angular matter "space waves" (quantum force) of each quark collide, as they're mutual gravitational fields (quantum gravity), pull them together. But once again we must apply Newton's Third Law of motion to this situation. The (action) is angularly driven matter waves of space colliding with each other and the waves are of course, connected to the quarks that generate them. This reflects some of the force contained in the interaction back at the quarks involved in this action (reaction). Note: this same process, when related to nucleon interaction, explains why the nucleus has less total mass than the sum mass of all its constituent mass parts.

Since the angular matter waves are basically generated by particles of light going in orbits within the quarks, and light maintains one speed, this causes angular motional interactions between the quarks, and the quark-triad (proton) as a whole unit will then rotate. This is why protons and neutrons, when freed from the nuclei they make up, have detectable properties of spin.

Note again, that by taking two identical quarks, both either left handedly or right handedly oriented and turning one quark upside down 180^0, they will come together and annihilate each other. This, I believe, is why nuclei always contain approx. the same number of neutrons as protons. Else, all protons would have eventually annihilated each other. Glashow said, "If quarks makeup protons, the proton will eventually decay into electrons and positrons, which will combine to form photons (annihilation), and the universe will end as it began, with light. This model agrees with the end result relating to this postulate, though it shows the path taken, to be somewhat different.

Up to this point we have addressed the mass particle made from the mixing of heavy and intermediate quarks, the proton. We must however address one other characteristic the nucleus possesses before we take on the electrons. What I speak of here is the addition of the particle which allows nuclei to form into long lived entities such as we have now in nature around us, namely the neutron. The neutron is slightly more massive than the proton and also has the strange ability of being what we call neutral. The mechanism that allows the neutron to be considered neutral, is a quark that was formed late in the mass particle creation period after the Big Bang.

Toward the end of the mass particle creation period, light (kinetic energy) undergoing this "piling up" process which produced the heavy and intermediate quarks was becoming less available. The quarks created during this later period we will call the weak quarks and they come in two configurations, weak and weak/neutral. The weak quarks all have their particles of light moving in the same direction within their orbits and go by the name of electrons. But the weak/neutral quark has approx. half of the particles of light making it up, going in the opposite direction from each other. This is possible by extending Pauli's Exclusion Principle so that, as long as the kinetic waves generated by the particles of light do not interact with each other head on, they can both exist within the same quark. We now have a quark simultaneously possessing duel matter waves with opposite angular direction. This would allow these neutral quarks to go through an electromagnetic field in a more or less straight line. The standard model shows one particle that I believe might be a neutral quark and it is called a "neutrino". Stay aware that all elementary particles are simply quarks with different amounts of light in orbit within them.

If we combine one of these neutral quarks (neutrino) with an electron, this neutral particle (nulectron) has the ability to approach a proton, be kinetically caught up in the protons gravitational field and actually form an orbit around it. This combination will create a particle with a neutral charge (neutron). In addition, we now have an explanation for the extra mass that the neutron possesses, and also for why the neutron can go through an electromagnetic field in, more or less, a straight line. An electron, with a neutrino in orbit around it, generates a dual (neutral) force effect in relation to its matter wave. This allowed it, later in the mass particle creation period, to be drawn into an orbit around the protons. And the neutrino, orbiting the outside of the electron, which together, orbits the outside of the proton, has a neutral force effect upon the other protons. This is what saves nuclei from annihilation and also allows for atoms with many more nucleons to form.

In addition, this also explains why this nulectron can migrate to another proton and then allows this other proton to become a neutron. This action is not necessarily a normal process for the nulectron but can happen during hi-energy events such as in a collider. I hope it is recognized that this approach solves several phenomena connected with the nucleus. You might also say that this Nulectron, under certain circumstances can migrate between protons.

I'm sure some have noticed I have just described the neutrino as a weak/neutral quark. In my investigating the logical sequence that mass particle creation would have gone through, and combining this with the standard model, I arrived at the following conclusions. Any particle with a value of mass has one or more quarks within it. Quarks generate 99.999% of the mass and gravitational fields we see in nature around us, with some small amount that results, from electrons moving around their orbits within the atom itself.

Only protons and neutrons have heavy quarks in their makeup. The heavy quarks were formed early in the mass particle creation period, but then came the production of the intermediate quarks later within this creation period. Because of the much larger gravitational field that a heavy quark's kinetic energy generates, it is allowed to trap and hold two of the intermediates. Keep in mind here that the intermediates also contribute some of the quantum gravity necessary for this configuration to evolve. The weak and neutral quarks have yet to be formed so we can discount their interactions at this point.

The ultimate configuration has the heavy quark, roughly in the middle of two intermediate quarks, and all three maintain the same interactive axis orientation within the space around them. This configuration still allows the intermediates to move around the heavy quark as they interact with each other and the triad as a whole can revolve. The matter waves (quantum force) being generated by the quarks keep the quarks from physically joining each other while the quantum gravity keeps the quarks in a close order combination with each other. This, scientists of year 3000 AD is how protons were formed.

As I said previously, as particles of "piled up" light, becomes more scarce, a third type of quark is formed and these I call the weak quarks. These quarks have nearly the least amount of particles of light, in orbits within them, allowed. The neutral quark (neutrino) has the least. High-energy (accelerator) collisions can cause the quarks within the protons and neutrons to decay into lighter, less dense particles of mass with short lives. Remember, within this model, all mass particles are made from quarks, and photons are "released", not created, as quarks decay.

We now have all the quarks for this model accounted for and I describe them in just four configurations; Heavy, Intermediate, Light and Neutral. Note: the Light quark is the electron and the Neutral quark is the neutrino.

My Standard model is as follows:

Quarks- heavy (H), intermediate (I), light (L), neutral (N)

Proton - (HI2) Particle possessing one heavy and two intermediate quarks

Electron- (L) Particle possessing one light quark

Neutrino- (N) Particle possessing one low mass neutral quark

Nulectron- (NL) Particle possessing one neutrino and one electron

Nupositron-(NP) Particle possessing one neutrino and one positron

Neutron- {(HI2) (NL or NP)} Neutral particle possessing a proton and one nulectron or one nupositron

Photons- output of Big Bang, quark decay and atom emissions

Universal Force	= Pressure (Generator of all other forces)
Quantum Force (mass)	= Quark matter waves (angular)
Inertia/gyre	= Effect of orbiting kinetic energy within quarks
Gravity	= Consequence of expanded space (guiding path of least resistance)
Electromagnetism	= Reactive space resulting from mainly electron motion

*I place gravity with the forces when in fact gravity is not an active force at all but only an inertial guiding path of least resistance within radiated space itself.

With the exception of pressure, all forces within this model are generated by space's reaction, to particles of light, in motion through it. In addition; no mass particles or forces are possible without kinetic energy in motion. Believe it or not, and as for you scientists in the 31st century, top this work if you can!!!

This model of the atom, was put together under the requirements that all mass objects are made from combinations of quarks and when quarks decay the result is photons. I basically try to use the Gell-Mann idea of quarks but do not address such things as up, down, color, charm, or strangeness, for in truth, I did not completely understand these properties, with the exception of force symmetry conservation. In addressing symmetry, I see it as the balance of forces, within a system composed of various quark particles. Without force symmetry balance, nuclei and atom creation could not have progressed as it did. The universe did not count as it generated all the bodies of matter within it. Basically, if the interacting forces developed a balanced configuration, a particle or atom was born.

I see the weak nuclear force, not as a force, but as a result of force "symmetry" imbalance within the atoms of the elements above iron (FE), in the periodic table. As I have understood it, the weak force was theorized to explain the emissions (alpha, beta, gamma) of radioactive matter. If we were to place a hydrogen atom and a uranium atom far out in space away from any massive objects such as stars and galaxies and given enough time, we would see two different types of decay. The hydrogen atoms' quarks would continue to radiate (decay) until after, tens of billions of years, the hydrogen atom would be nothing but photons, all moving off thru space in various directions. The uranium atom ends up the same as the hydrogen atom but it undergoes a different kind of decay along the way.

All the atoms making up the elements up to iron (26) exist in what we might consider a semi-stable atomic condition (neutron/proton balance). The elements above iron lose this balance and as such, can experience major decay. By this I mean, the elements above iron, not only experience quark radiation emissions, but can also emit whole protons, neutrons, electrons and, high frequency photons with total quark decay (gamma). This is due to a proton/neutron force imbalance, which first causes accelerated decay of the light/neutral quarks making up the nulectrons.

The extra nucleons that join to create the elements above iron are not in a restrained configuration such as the nucleons that joined to create iron and below. This allows the nucleons making up the elements above 26 to move around the nucleus in addition to their individual spinning. Since the combination of two protons and two neutrons is the most stable set nucleons can exist in, we create a situation where this "alpha particle" is in haphazard motion around the stable portion of the nucleus. The component that decays first, in this instance, is the part of the neutron that makes it neutral (nulectron). And within the nulectron the neutrino is of course the first to completely decay. This, my friends is "fatal" entropy in action. If man can survive long enough, there will come a time when elements above iron in the periodic table will become scarce as hen's teeth. He will then have to find another Big Bang out there somewhere.

As I postulated before; an electron joins with a neutrino quark (LN) and then this nulectron, joins with a proton to create a neutron. So we basically have a neutrino orbiting an electron, which in turn is orbiting a proton. This puts this outer orbiting neutrino in a configuration where it is in constant conflict with the forces around the stable nucleons. This causes the neutrino to radiate at a magnitude above its normal emission rate and when enough energy is lost by the neutrino, it experiences accelerated decay. This now destroys the electron's ability to stay close to the proton and it is eventually ejected as a beta particle. This in turn now creates the situation where we have protons without a neutral barrier between them. When alpha particles decay in this way and under these conditions, this can cause "whole" alpha particles in the immediate area around where the alpha particle decayed, to be ejected as well. Note: if a proton interacts with a neutron with enough force it can sometimes steal the nulectron (LN) away from the neutron.

Up to this point we have seen space matter, under the force of pressure (Big Bang), being transformed into light, then under the stern guidance of the gravitational fields become quarks (particle containing light moving in orbits), which will eventually decay into photons (light herded together and moving in linear directions). And as these particles of light move outward beyond the gravitational domain of all the matter created after the Big Bang they will be reabsorbed by the much denser fabric of space in what I call, outer-outer space. Somewhere out beyond this outer-outer space probably exists mass/energy remnants of other Big Bangs from times gone by.

Newton recognized the need to put laws around the effects of matter in motion. The effects had already been pointed out by the analysis of such men as Kepler and Galileo. Newton seen that all mass objects had a tendency to stay in motion when in motion and also a tendency to stay at rest when at rest.

Newton also seen the need to apply an equation to the effect (force), one mass object can have on another mass object. Hence the equation $F = ma$. I see this equation as follows, "The potential effect, one mass object can apply to another mass object, is directly related to the magnitude of the mass object's, quark matter waves, in motion, through space, at some velocity. Of course the mass of both objects must be inserted within the equation. The main point to be made here is that this force is not an illusion, but can be described as a kinetic manipulation of the space around an object of mass. Force is ultimately created by particles of light in motion, hence, all magnitudes of force will have, at their foundations, Planck's constant.

As an example, let us use the electron. The particles of light, in orbit within the electrons (quarks), are causing space to generate angular kinetic waves (matter waves) around the quarks which drives the space around the quarks. This angularly driven space is what gives the electron's (quarks) their mass and under this situation we can see that mass is nothing more than an angular spatial force which resides in the "driven " waves of physical space surrounding the quarks.

Within this work, when we put this electron in motion we will see that the mass (angular force) starts decreasing the faster the electron moves. But the electron's total force is not disappearing, only repositioning itself. The particles of light within the quarks, when the electron was stationary with reference to the point in space where the Big Bang occurred **(^)**, were making their orbits at speed c, but when the electron is put into motion, the particles of light in the quarks, must now share their "orbital" speed within their orbits, with the speed of the electron, as it moves through space. This causes the frequency for the orbit period, of the particles of light within the electron, to become longer. The particles of light did not slow down, but they must now share their speed with the orbits they are making within the electrons and also the direction the quarks are going thru space. One direction being the orbital track within the quarks and the other being the direction through space the electron is moving. This is the same process I describe elsewhere within this work.

The end result is, the quarks cannot increase the driving of the space around them as much, with the electron in motion, as they can when the electron is stationary. This in actually, is a loss of angular force (mass). But even though the force of mass is decreasing, it is being transferred, due to motion, out ahead of the electron as a linear kinetic wave (deBroglie's pilot waves). 3rd law again. In addition; as the electron is put in motion it foreshortens and becomes much more compact which causes some to believe "mistakenly" the electron experiences an increase in mass.

And this is all dependent upon the velocity of the electron. In 1900 a man by the name of Lenard performed an experiment to confirm that the photoelectric particles were actually electrons. An interesting side-result is that he also showed that the photoelectric particles (electrons) can possess different magnitudes of kinetic energy. Within this model, the matter waves (quantum force) of an electron (quark) is a result of the internal actions of the kinetic energy within it. This agrees with the postulate that the mass force within an electron is converted as the electron is put into motion. It also explains why a larger voltage produces a larger electromagnetic field. In addition, we can now understand why it takes a charge "electrons in motion" to generate an electromagnetic field. I have no doubt that protons and neutrons in motion can also generate electromagnetic fields. Remember that electrons (particles of mass) in motion generate electromagnetic fields. Protons and neutrons are also particles of mass hence, if you align them and put them in motion an electromagnetic field will be generated. A side note needs to be inserted at this point.

At present, orbital motion within atoms is seen to be the equivalent of acceleration and this assumption was used to show that Rutherford's atom could not be stable in the classical sense, and would, in fact, collapse. This assumption comes from the view that an electron, when in motion in a straight line at a constant speed, will not radiate. But when force is applied to the electron, which results in its taking up an orbit (velocity), it must now be considered to be under the influence of what is called acceleration. And an electron is known to emit light when under the influence of an accelerative force. The key to this misconception is the ignorance of what gravity really is and the misuse of the term "accelerative force".

Science presently sees gravity as a force and when it is applied to an electron as it obits the nucleus this is seen as an acceleration for the electron. Under this reasoning, Rutherford's electrons should be continuously radiating, which would eventually cause his atom to collapse. But it has been shown that an electron in orbit does not continuously radiate and a postulate was put forth, using Planck's constant to create energy shells, in an attempt to explain this phenomenon. Take note that black holes evidently cannot capture everything. If they truly sucked in the theorized gravitons then how would they work as described. The scientific community at present sees gravitons as a radiating particle yet they somehow are able to escape black hole death. This is a major weakness in black hole theory and must eventually be addressed.

ATOM CREATION

I will now describe the actions taking place in order to allow atom creation. Within this model, the reason electrons do not radiate while in orbit around the nuclei, is because no accelerative force is being applied to them. Gravitational fields offer themselves to the electrons in order to allow them to take up orbits, so gravity, we must now realize, is not an active force but is an inertial guiding path of least resistance. This is why gravity itself cannot accelerate the electrons within atoms, only offer an inertial guiding path of least resistance. In other words; the electrons make themselves fall or accelerate. In a semiconductor, electrons can be made to jump the gap from n to p and they emit a quantum of light as they do (LEDs). The jump was due to external forces being applied to these electrons, which caused the electrons to undergo the force known as acceleration. So we can say that force being applied to an electron from an external source is what causes it to radiate. Yet within an atom this seemingly same process relating to electrons does not get the same result. Only an inertial guiding path of least resistance can solve for this mysterious phenomenon.

This same "force" that Newton describes in the 1600s is the same force we see riding just ahead of the particle we have come to know as the electron (DeBroglie's pilot wave). Angular force when related to the immediate area surrounding electrons, generates what we call mass. So; the light spit out of the Big Bang generates a large gravitational field while also piling up within this gravitational field to produce quarks. The point to be made is, an object's force must be treated the same as that object's energy. If we look at the system (universe) as a whole, all we are dealing with is mass at rest and mass in motion. When an object is at rest it still possesses a rest energy or rest force (mass). And an object's force (mass) is directly related to its energy. This says, when an object emits or radiates, it will experience a slight loss of energy, and also force (mass). If we put the two equations $E = mc^2$ and $F = ma$ under analysis we can see that E = mass times motion and F = mass times motion. And these values are directly divisible by a factor we know as Planck's constant. And Planck's constant leads us, of course, to light. And this leads to the inescapable conclusion that energy is the result of particles of light in motion and force is space's reaction to particles of light in motion. Hence, E=F is nothing more than the particle/wave duality showing its face in nature around us. (E = F) = (action = reaction). GOD: I don't know much--- but I know I love you--- and that may be--- all I need to know!

A photon in motion at the speed of light has zero mass, but has an amount of energy equal to the number of particles of light within the makeup of the photon, times Planck's constant.

And just as the total energy of this photon is divisible by the Planck units of energy tied to the individual particles of light making up the photon, so is the force of the photon. This says, force is ultimately, the effect of energy in motion. In short, the energy of light is the light particle plus its motion, and the force of light is space's reaction to this kinetic energy and this force shows itself as the kinetic wave that rides just ahead of the particles of light.

When a particle is at rest, its force is the angular matter waves (mass) surrounding it. When that particle is put in motion, its force begins a conversion process into a kinetic wave riding just ahead of the object, and this force grows with velocity. This means, as the "linear" kinetic wave force increases, the "angular" matter wave force (mass) must decrease, and the decrease in angular matter wave strength is also known as a loss of mass. As shown before, the particles of light, traveling in orbits to make up the quarks, give the quarks their mass, as a result of the angular matter waves they generate around the quarks. This is F=ma carried to its ultimate conclusion and shows that mass is an angular spatial force and all forces, with light being the sole exception, is a spatial effect of mass in motion and takes place within space itself.

The alternative to this approach would be to hold mass constant or as is presently believed (erroneously I might add), to allow it to increase with motion. This approach I found unacceptable as this would equate to an increase in energy for the object put into motion and this is not allowed due to the fact that mass is not energy but is a side effect (force) of energy in motion. Mass is simply the reaction by space, to particles of energy, traveling in orbits within it.

Let us now jump from Newton to Maxwell. Maxwell took the work of Faraday, who intimately studied the phenomenon we call electromagnetism, and Maxwell devised a mathematical tool to describe what Faraday experimentally showed. Maxwell weaved a tapestry of field equations relating to electromagnetism that still stands today as one of the few all time greatest works, in mathematics.

But timing is everything, someone said. Maxwell's equations are built upon the idea of a physical space, which he used to dampen and restrict the forces of electromagnetism (no infinities). He did this work around 1875. In the 1880s Michelson and Morley performed a series of experiments to validate the existence of the aether (physical space). Their attempts failed and the result is now seen as a case where something (space) is now considered a void (nothing). Maxwell could never have constructed the field equations without his belief in a physical space.

Imagine someone suggesting to Maxwell that he should imagine a force carrying particle (light) that is radiated by the electromagnet, which, when its collides with a ferrous object, causes this object to be put into motion in the opposite direction, of the direction in which the force was applied. And I won't even mention a certain mono-poled particle some believe might exist, that performs the same action in relation to gravity.

Maxwell placed his fields within physical space and "disturbed" space actually was his field. Maxwell's field can be described as an area of space to which he gave reactive values, in accordance with Newton's 3rd Law of motion, and these reactive values were precisely defined at every point within this field. All his forces diminished in accordance with distance, through this reactive field (space).

With all this said, we are now ready to put the atom in motion. In my analysis of the mass particle creation period after the Big Bang, I say again, the quark building stage was the first and only major stage for all creation. And the most significant influence on stage at this point was simply space's reaction to kinetic energy flowing thru it; the gravitational fields.

We have now set the stage for joining the electrons with the nuclei. The first thing we must address is why does the atom collect its electrons, only in certain specified numbers and orbits? Using the assumption I postulated above, we must conclude that the makeup of the nuclei, have a direct connection relating to the number of electrons it will surround itself with. So we now analyze how this process might work, and we immediately recognize the footprints left by forces at work. Those who see charge as the controlling force here I believe, need to investigate further. I cannot believe that the nucleus contains over 90% of the atoms' mass/energy but can only hold or maintain an equal charge with reference to the surrounding electron charges.

Science at present sees four forces at work; weak, strong, electromagnetic and gravity. My model uses quantum gravity, quantum force and electromagnetism with which to work this problem. Because we are seeing force interaction at the atomic level, we can rule out quantum force and quantum gravity on an individual quark basis, but when quantum force and quantum gravity are taken collectively for the nucleus as a whole, this is a different situation. Nuclear force (nucleus matter waves) dies out quickly as it is extended beyond the nucleus, but it does show some amount of force into this area outside the nucleus. We have now set our minimum distance for the first shell or ground state that the electron is allowed to occupy.

The electrons themselves, while under the extra effect of the gravitational field surrounding the nucleus, control the rest of the process with one exception. How does the nucleus capture the electrons in the first place? This is accomplished by the collective strength of all the quantum gravity (nuclear gravity) generated by the quarks making up the protons and neutrons, within the nucleus. So now we have the nucleus with a small force field (mass) around it and a large gravitational domain, per the size of the atom, that extends well out into the area around the nucleus. Note; the electrons within the atoms also add their gravitational fields to the fray.

Let us take a hydrogen nucleus as an example. The proton, at some time in the past, existed without an electron in an orbit around it. This was of course; rectified when electrons eventually came on stage and the first electron came wandering by. The proton's nuclear gravitational field offered an inertial guiding path of least resistance to the electron, and in it came. Since the inertial guiding path of least resistance provides an easy pathway for the electron to the proton, that is the direction the electron takes. But before the electron can physically merge with the proton, it encounters a repulsive force, whose strength increases very quickly, the closer the electron approaches the proton. So the electron ends up in an orbit around the proton that is governed by its interaction between nuclear matter waves and nuclear gravity. We have now constructed the atomic configuration associated with the hydrogen atom, but we still have one problem to confront.

Why is there only one electron in this first shell or ground state? As we discussed before, the electron, when put into motion causes space to react by generating a kinetic wave just ahead of the electron (DeBroglie's pilot wave) which then results in a standing wave stretching out behind. This standing wave I speak of here will leave a trail (disturbed space) of considerable length as it orbits around the nucleus. This is the reason other electrons will find it physically impossible to take up residence in the same orbit (ground state) as the first electron does. This is how Hydrogen is produced.

Now let us take two protons and add the two neutrons that will allow us to combine the two protons close together. Once this nucleus combination is set up, along will come an electron and it will be enticed by the gravitational field surrounding the nucleus into taking up an orbit in close to this nucleus. No doubt that in the early mass creation period right after the Big Bang occurred atoms existed in the configuration just described but that configuration would not have lasted for long. You see; the matter waves emanating from this nucleus (two protons-two neutrons) are much stronger than the matter waves emanating from a hydrogen nucleus (one proton) hence, the ground state for this nucleus will be farther out and this configuration will allow another electron to enter into the same ground state orbit, though not the same exact location, as the first electron is occupying. This is how Helium is produced.

An electron in motion causes space to generate a kinetic wave just ahead of the electron and I call this de Broglie's pilot wave. Note that causing the electron to change direction back and forth (alternating current) in a conductive wire can provide a shocking sensation to anyone who touches it while in contact with a grounded source. To repeat what was just discussed above; A side effect of this kinetic wave is what we call the standing wave. As the electron orbits and evolves around the proton it disturbs the space within this orbit again and again and this disturbance becomes what is known as a standing wave. The standing wave (physically disturbed space) is a reactive force and when another electron is drawn in by the proton's gravitational field, it runs into this standing wave and is repulsed. This is why a hydrogen atom has only one electron. This is also, I believe, the base factor for truly understanding Pauli's exclusion principle. Note here that photons can be captured by atoms but only between the electron shells.

Now we will investigate how the atoms can somehow end up with more than one electron in their ground state. As you add more and more protons and neutrons the nucleus becomes larger in diameter and also generates a much stronger spatial matter wave field and spatial gravitational field around itself. This causes the first electron captured to take a ground state far enough away from the nucleus so that its standing wave is unable to cover all the possible approaches to the nucleus. So now another electron can be captured and it is allowed to fall into the same orbital as the first electron. But the standing wave of this second electron, when combined with the standing wave of the first electron disallows any other electrons to fall into this ground state. This, I believe, is the base mechanical factor behind what we call Pauli's Exclusion Principle. (See Figure 1.) Note that Figure 1. (two dimensional) shows an abbreviated description of the Helium atom for the standing waves in reality extends well beyond just one orbit and direction.

FIGURE 1.

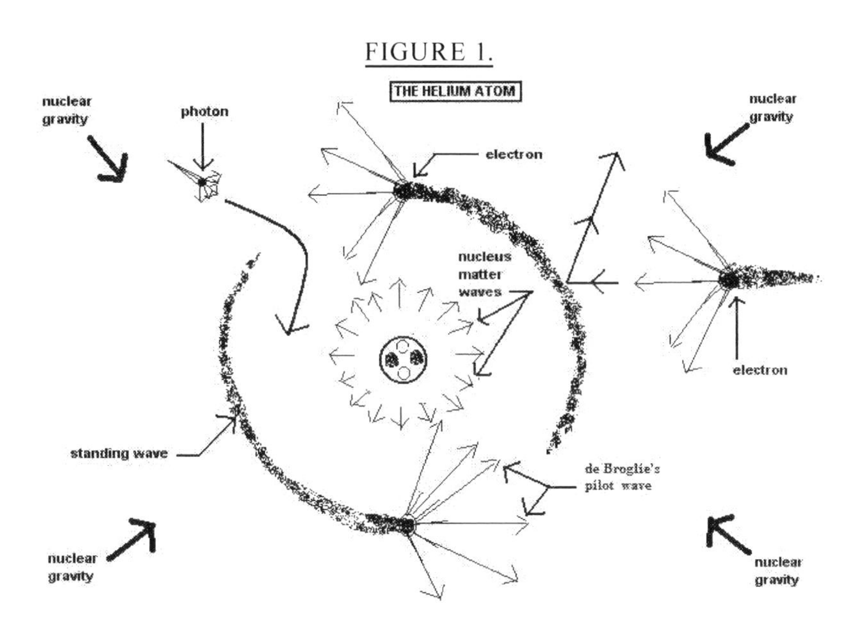

So basically we can say that the ground-state is determined as a result of the interaction between the nucleus's matter waves and its nuclear gravitational field along with the electron's trailing standing wave. The nucleus's associated gravitational field provides an inertial guiding path of least resistance which is more than sufficient to allow more than two electrons into the ground state but the reason the ground state can, at most, contain only two electrons is because of a combination of repelling forces between the nucleus's matter waves and the two ground state electron's standing waves. With the exception of the ground state, the total number and positioning of electrons an atom will eventually end up with is determined by the gravitational field generated by the electrons, protons and neutrons the nucleus possesses along with what the standing waves will allow.

Again; the number of electrons allowed in the ground state is determined by the interactions between the nuclear gravitational field, the nuclear matter waves and the electron's standing waves. All the other shells and the number of electrons within these shells are determined by interaction between nuclear gravity and the electron's standing waves plus the electron's matter waves. The equation needed to describe all this must, as a minimum, show the interactions between matter waves, gravity and standing waves.

It goes like this; when an electron approaches an atom that has lost some of its electrons, the approaching electron finds it easier, due to the atom's gravitational field, to travel in a direction towards the atom. So in the electron goes, but it then finds that it is being repulsed from taking up an orbit within a filled up shell, by the standing waves generated by the electrons already in the filled up shell. Then the electrons in the filled up shell come around and their matter waves force the new electron out away from the filled up shell. This action then allows the new electron to take up residency in a higher shell where standing waves are such that an opening is available. It's that simple.

During the period right after the Big Bang occurred, atomic makeup and molecular creation are determined by pure chance under the combinations allowed by force symmetry. The alternative would be only one type of atom, making up only one type of molecule, making up only one element. Chemistry gives a good description of what happens after this stage so we will not proceed farther along this line of investigation. Let us now return to the standing wave the electrons produce as they travel around the nucleus.

To reiterate: if the mass (angular matter waves) of an electron is nothing more than space reacting to the particles of light orbiting within the electron in either left or right handed direction, then by use of the 3rd Law, we must recognize that space itself will have to further react to this action being taken on it when the electron is put in motion. And it does. That reaction turns out to be the replenishing of the space the electron drove ahead of it. So as the electron moves along, and drives the space ahead of it, space is forced to replenishing itself after the electron moves through. This action generates what I will call a trailing spatial wave or better known as a standing wave.

Recalling what was shown before; quarks make up all bodies containing mass. If we apply this to the atom as a whole, we find that the electrons are somewhat driving space away from the atom, as a whole. This is what I spoke of previously when I said that electron motion also contributed some small value to the mass force and gravitational field around the atom.

The process that was shown here for the hydrogen atom applies to all the other atoms also, and in the same manner. The number of shells or orbits, an atom can control, is directly dependent upon the number of protons and neutrons, the nucleus contains. The number of electrons within each shell is controlled by the trailing (standing waves) connected with the electrons themselves. Again; this is why only two electrons are allowed in the first shell. And the matter waves plus the standing waves of these two electrons along with the gravitational field surrounding the nucleus will dictate where the next shell will be formed.

In addition, with reference to atoms of a higher element, if radiation is applied to the atom as a whole, the radiation is not absorbed by electrons as many believe, but instead is herded by the matter waves and standing waves into the areas between the shells. Note: when radiation is applied to an atom it is believed presently, that the electrons within the atom somehow absorb the applied radiation. But that is an erroneous conclusion. The electrons, as was explained before, possess matter waves which are constantly being angularly generated into the space surrounding the electron and any radiation that tries to penetrate these electron matter waves will find itself rebuffed. Now a situation is arrived at where the kinetic wave property of the radiation within the area between the shells combines its kinetic force with the force of the matter waves belonging to the electrons within the next inner shell and this then causes the electrons within the next outer shell to be forced into a higher orbit, to be calculated by use of Planck's constant, farther away from the nucleus. I believe this same action can also cause the outer electrons to be forced into a higher orbits also, such as, I believe, happens with phosphorus. Keep in mind that the terms electron and quark are synonymous within this model.

This model shows that within the atom as a whole, the nuclear gravitational field and the electrons themselves dictate the number of electrons within a shell and the distance from the nucleus, these shells will be. The total possible electrons an atom can contain is dictated by the number of neutrons and protons within the nucleus along with the strength of the gravitational field the nucleus is generating. Now for the supposed "Quantum Leap"

If we look within the electron and analyze the quantum force (orbiting kinetic energy) at work there, I see no way for an applied photon to breach the kinetic force residing there. De Broglie showed that electrons generate waves (forces) around themselves and since the radiation that is absorbed, have no value of mass, I can't imagine how they could get inside the electron. We must also remember, it is not the number of photons but the frequency of these photons that act upon the electrons. Also, if all electrons are deemed to be identical then why would some absorb different frequencies than others?

In addition, this approach must still be able to explain the electron's ability to absorb radiation and then release it very slowly and at a lower frequency, as in what is known as phosphorescence. This would mean that the electron would absorb the radiation and then horde it somehow to be released a little at a time. This puts an added equation into the mix, which would say that electrons have different emission rates for radiation. I see no way an individual electron can store the huge amount of radiation needed in order to emit it slowly back over the period of several minutes or hours. When we consider phosphorus with only five valence electrons and the number of photons that are emitted over time, I cannot imagine the phosphorus atom, as a whole, being able to store and regulate emissions such as this.

Put all this together and we can now see and understand the mechanical cause for the electrons' jumps. Finally, after the electrons have moved into a higher orbit the applied radiation finds it much easier to escape and the radiation is forced out of the atom and the electrons drop back to the orbits they were in before. The magnitude of the electron jumps are directly related to the energy of the radiation the atom absorbed, and the kinetic energy of the radiation is directly tied to Planck's constant. Hence, all electron jumps will show magnitudes with multiples of Planck's constant as a base factor. In addition, because the electrons are actually quarks, which are made of orbiting particles of kinetic energy (light) which also generate matter waves (mass), the orbit shells should all be graduated in distances that reflect these properties as factors also.

In addressing phosphorescence, I will offer the assumption that the element Phosphorus (atomic Number 15) has an electron configuration, which allows the atom to absorb radiation of several different frequencies, and then direct this radiation into the areas between the somewhat stable shells below the five valence electron's outer shell. So in effect, some of the applied radiation also breaches the gaps between the standing waves of the eight electrons in the next shell. This would explain why it takes so much time for all the applied radiation to be re-emitted. The electron's matter waves and standing waves allowing interactions with the radiation would be the main cause for this.

This also says that the radiation re-emitted within the first few minutes after absorption should have frequency magnitudes greater than the photons re-emitted later on. Using the herd analogy, we can assume that more horses (particles of light) can escape at one time, from the five big gates (outer shell) than from the eight smaller gates (next inner shell). Again; due to the nucleus matter waves residing just inside the first shell outside the nucleus, I don't see any photons breaching inward thru this barrier.

Using this approach, if we bombard phosphorus with light of a specified frequency and then measure the frequency of the resultant emissions, both immediately and later on, we can then calculate approximate values to be used for investigating the interior of the atom itself. Note: atoms are the result of balanced force symmetry within and between quarks (protons neutrons, electrons and neutrinos). It's really just that simple. The rest of this model mostly agrees with the work being done by chemists, so I will end this particular avenue of work at this point.

Be aware that many within the scientific community will say that my work here is but classical in its approach and it rejects the Quantum Mechanical Theory. I say this is absolutely not the case for this work, at its base, is exactly in line with what Quantum Mechanical Theory should be. You readers, both now and in the future, be the judges of this particular assessment.

THE ATOM

I realize I am being redundant with what I'm about to present but bear with me. In an effort to better let you understand my ideas relating to atoms, we need to return to the event known as the Big Bang, as this is the direct cause for the configurations that atoms take up at present. The Big Bang is simply space erupting within itself, due to intense pressure being applied to it. I speak of this process in an earlier portion of the work I have assembled here.

The direct output of the Big Bang was kinetic energy in the form of radiation and this immense outpouring caused a huge kinetic wave of kinetic energy in the space around where the Big Bang occurred. This resulted in the space around this area being greatly expanded (gravitational field) and this in turn would allow the radiation being ejected, to attain a speed beyond what we recognize, as the "speed of light" or c. After some time, the leading edge of all this ejected light starts meeting up with denser space as the Big Bang's kinetic wave becomes dampened the farther it travels away from the area in space where the Big Bang occurred. This sets up a situation where the leading edge of all this ejected light begins slowing back down to c as it continues its expansion into this denser (more normal) space. Take note; excluding the pressure being applied to space at this point we have only one force (radiation pressure) present which is acting to generate a huge gravitational field within space itself. But meanwhile, the light coming behind the initial radiation being ejected, still running above c, starts overrunning the slower, leading edge of ejected light. This is the beginning of the period after the Big Bang, in which mass particle (quark) creation occurs.

Light, left to itself, will travel in a straight or linear direction as we see in the particle we know as the photon. But the situation I am now describing results in a deviation from this norm. As the trailing edge of the ejected light overruns the leading edge, this causes the individual particles of light to deviate from their normal linear motion through space. This "piling up" causes the individual particles of kinetic light to take up orbital (non-linear) paths around each other. I must leave this line of thought for a moment in order to explain another event that is also occurring at this same time, which will directly affect the next step of this mass particle creation period.

There is a law known as Newton's 3rd Law of motion and it states, "for every action, there is created, an equal and opposite reaction, in direct response, to the action". I will add to this the following, "the reaction, generated by the action, always manifests itself in the area immediately around, where the action occurred". An example of this can be shown, by describing the pressure being applied to space as the action, and the Big Bang as the reaction.

But we need to also realize that this reaction can also be seen as another action, which can generate another reaction (high-energy physics). What I speak of here though, is that the original pressure that caused all of these events did not simply disappear. The radiation output from the Big Bang was the reaction and shows up as radiation traveling at a high rate of speed through the space around it. But in this instance, it is still under, and will continue to be under, the influence of the original action (cause), which is pressure. Pressure rules and will continue to rule throughout.

The universe was, is, and always will be, under extreme pressure. Pressure, within the matter making up the fabric of space, is what generated the Big Bang. In addition; pressure on the individual particles of light ejected from the Big Bang, produce all the matter (quarks) in nature around us. The effect of this constant pressure/motion process is what we call entropy. This same pressure is also responsible for the speed that light attains. What I am postulating here is; the speed with which light travels, is directly connected to the pressure applied by the space immediately surrounding the individual particles of light. Big Bangs are the result of pressure being applied within different areas of space itself.

This work's Standard Model basically uses a configuration containing particles of space, which under pressure, are put into motion (Big Bang), and becomes what we call radiation or light. This light, while still under the influence of the Big Bang, creates a mass particle we know as quarks. Quarks come in various mass densities depending upon what part of the mass particle creation period they were formed in. All objects of mass are in fact, either a quark or a combination of quarks.

I need to make it clear that quarks are basically, particles of light gravitationally locked into orbits around each other and if I should speak of mass to energy conversion, I simply mean that some the particles of light making up the quarks are released from within them. As these quarks decay, the particles of light making them up are released in various directions to become what we call photons. The frequencies of photons emitted from an atom are directly tied to the electrons' standing wave configurations within the atom (Pauli's exclusion principle). Photons within this model are made from different amounts, of particles of light and these particles of light are all moving in a straight line as they travel through space. This is why they have no mass. It takes particles of light traveling in circles (quarks) to produce the angular spinning of space (quantum force) necessary to produce what we call mass. Sorry about your Higgs but I just don't need it. I have found that kinetic energy is responsible for the production of all mass particles and forces within nature around us.

This process has two direct products called gravity (expanded space) and angular matter waves (quantum force). Basically we can say that particles of light in motion, through space, make up all of nature around us. The reaction by space, to particles of light moving through it, is what we have come to know as force. The exception to this is gravity, which is an inertial guiding path of least resistance (expanded space) and cannot be treated as a positive force. Within this model, the fabric of space can be contracted, expanded, driven and spun. I'm not saying that it is impossible to fold space, but I see no way, given our present level of technology, for objects of mass, to affect this result. Also, when I speak of contracted space, I am referring to the effect entropy has on space, as it breaks down mass and returns its energy (light) back to space itself, which then puts pressure (contraction) on it.

Space is converted into kinetic energy during the Big Bang. A small part of this energy is converted, to what we call mass right after the Big Bang. Annihilation between quarks is the main reason a large amount of the ejected radiation from the Big Bang, did not go on to become mass.

Particles of mass are presently thought to possess a property of spin although no one seems to know just exactly how they spin. What is spinning is quarks (electrons and neutrinos) or combinations of quarks (protons and neutrons). These particles revolve within space itself, within and around, the atoms. Mass is caused by the particles of light traveling in their orbits within the quarks and driving the space ahead of them, which in effect, gives these quarks what I term, an angular force (quantum force). Quarks necessarily revolve within the space they reside within hence, this is the spin phenomenon they seek. The action mostly responsible for this spin is also the force that keeps quarks from coming together and annihilating each other (matter waves). This is why I call it the quantum force (DeBroglie's pilot waves or matter waves).

To reiterate: within this model entropy (pressure) acts constantly to reduce all particles of matter back down to just single kinetic particles of energy which are then reabsorbed by space thus setting up ultra intense pressure systems within space. Big Bangs (space rupturing under pressure) are the result of just such a design. Note that under an arrangement such as this, our Big Bang would be but a local event. Note also; this points to a conclusion which says the only true force acting around us in nature, is pressure. All the forces can be reduced to manifestations of pressure. These bits of information I am presenting here are for the science students of year 3000 in order to let them know that not everyone at my present time was unable to see beyond the obvious.

A good everyday analogy of the forces we assume to exist within mass would be to get a merry-go-round spinning and then try to step up on it. It immediately throws you spinning out into the yard. And depending upon which way the merry-go-round is spinning determines which way you will be thrown.

Quantum gravity (inertial guiding paths of least resistance) keeps the quarks bound to each other while the quantum force (angular quark matter waves), keep them from annihilating. Annihilation can occur however, due to the internal orientation between the quantum forces of two interacting quarks, and I have spoke of this elsewhere within this work.

Again; I need to make it clear that gravity cannot be treated as an active force, but is in fact, passive in character. The gravitational field is generated by the same mechanical action that generates the quantum force which is in fact kinetic energy. Again; the reason gravity cannot be considered as an active force is because it basically exists, without having any significant force characteristics tied to it. Gravity, unlike the forces, should be considered as a static, asymmetrical configuration of space, rather than an active particle generated physical force.

To use an example; take the keys out of your pocket, hold them straight out and then release them. Note that it seems as if the keys are immediately pulled downward by some mysterious force. This, in reality however, is not the case. I will show that in fact, the keys make themselves fall. (See Gravity on page 143)

At this particular point let me address a particularly worrisome subject you humans seem to overlook. In this year 2016 you are proposing efforts to travel to the planet mars and that is fine. But in order to travel beyond our solar system you must learn how to manipulate and master the periodic table (elements). Out in deep space you will run out of necessities such as the raw materials needed for repairs on you ships, the food, the water, the oxygen and fuel. I advise immediate concern in mastering this task for time is of the essence.

PART TWO – LIGHT

LIGHT

The GOD particle; a dual state particle and the basic particle making up our "local" portion of the universe and everything within it. Light, in its kinetic energy state, contracts and goes on to make up all of nature around us with the exception of space. In its rest state it expands and becomes part of the fabric making up space itself. In the foreword I posed the question; "Have you ever wondered how a particle, traveling at 186,000 miles per second, can hit something and rather than cause extensive damage, just glance off and go in another direction while never slowing down"? I investigated this phenomenon and came up with only one explanation that could account for this phenomenon. Light is pressured thru space by the fabric of space itself. This is why light fails to slow down even when interacting with bodies of mass. In addition; I need to make it clear that particles of light travel "thru" space while particles of mass travel "within" space. This will be addressed to a greater degree later within this work.

The Big Bang transforms the fabric of space (expanded particles of dormant energy in a state of rest) into kinetic energy (contracted particles of energy in motion), which then gravitationally interacts to produce what we know as mass-matter. Entropy eventually causes it to return to its rest state within space itself where it re-expands to provide the pressure needed to generate another Big Bang somewhere else in space. Note again; light particles are also the generator of all waves/forces. Because light in motion makes up all mass matter (quarks) within the universe, the speed associated with it is the rate of motion the universe uses as it moves into the future. Hence, the speed of light is the true standard for the universal or ultimate clock. It must be understood that with reference to the universe as a whole, everything happens at the speed of light for even though we might think we are standing still in the space around us the kinetic energy making us up is, in fact, in motion at the speed of light. This fact will become more obvious as we go along. Those of you in the 31rst century must realize the obstacles I've had to overcome to get this far in constructing a feasible configuration for the universe as a whole. It's very hard to push a theory when no one will even give it the light of day.

This model presents energy as the basic building block for the whole universe and you would expect characteristics of it to pop up in many theories within the scientific community. Think about the value 6.626×10^{-34} joule-second, and how many different places within theories you have seen it appear. This value is called Planck's constant and is directly tied to light (photons) which consists of numerous, individual, particles of light herded up together and traveling in mostly straight lines.

Einstein postulated that to accelerate an object to the speed of light, an infinite amount of energy would be needed. I disagree for I believe the Big Bang had enough energy to perform this feat (See the works of Gamow, Alpher, and Herman). The fact, that one can strike a match so easily to produce kinetic light, led me to only one conclusion. The kinetic energy (light) must have already existed within the matter making up that match. Note again and keep in mind; space pressure upon the individual particles of light is what gives light its motion and speed. It will eventually be shown that space's density determines the speed with which light travels thru it. I will be long gone by then but by this work you will know that I was here.

A basic particle of light, when traveling "thru" space and with its kinetic wave disturbing the space just in front of it (Newton's Third Law), basically excites (expands) the area of space it is traveling thru. This action (space reacting to the light traveling thru it) is the exact cause for the creation of gravitational fields.

Around 1900 and by use of what we call black body radiation, it was found that at high frequencies the energy of light falls off to zero as the frequency gets higher. But we must remember that measuring the number of oscillations passing a given point over time, is the process used to determine the frequency and energy of photons.

Within this model, the photon is seen as a particle made from numerous particles of light in a mostly linear grouping. The frequency of each photon is directly influenced by the number of light particles within the photon. At present, as a photon passes through a detector the oscillations are counted and the frequency and energy of the photon is determined. So we can say that, as each particle of light passes through the detector, its oscillation (wave) is measured.

But what does our detector see when the oscillations become so close together that all is seen by our detector is a continuous oscillation. Because it can no longer see the separation, it shows several oscillations as only one oscillation. Planck's constant was eventually used to address this problem even though Planck himself never believed his constant, in any way, could be connected to an actual characteristic of nature. He saw his own work as a mathematical deception.

The point to be made here is; as a particle of light within a photon becomes closer to the particle of light ahead of it (gamma), the oscillations (waves) start to merge. The point is finally reached to where an oscillation counter cannot distinguish any distance between these particles and their waves. At this juncture the counter will mis-register the value for the energy within the photons. Note that this result is possible only with a physical space in play. A particle of light in motion generates a frontal kinetic Planck wave of energy in the fabric of space immediately ahead of the particle. This is why we always find a particle and a wave when we investigate radiation.

When a quark radiates or decays, numerous particles of light are ejected in all directions at a certain frequency that depends on the exact nature of the decaying action. We can detect this decaying action and have come to know the part we detect as a photon. All the smaller kinetic waves associated with the individual particles of light combine to form a large kinetic wave that rides ahead of the photon (See Figure 2.)

Note: The pictorial of the photon below is only a close approximation and does not show its actual configuration.

FIGURE 2.

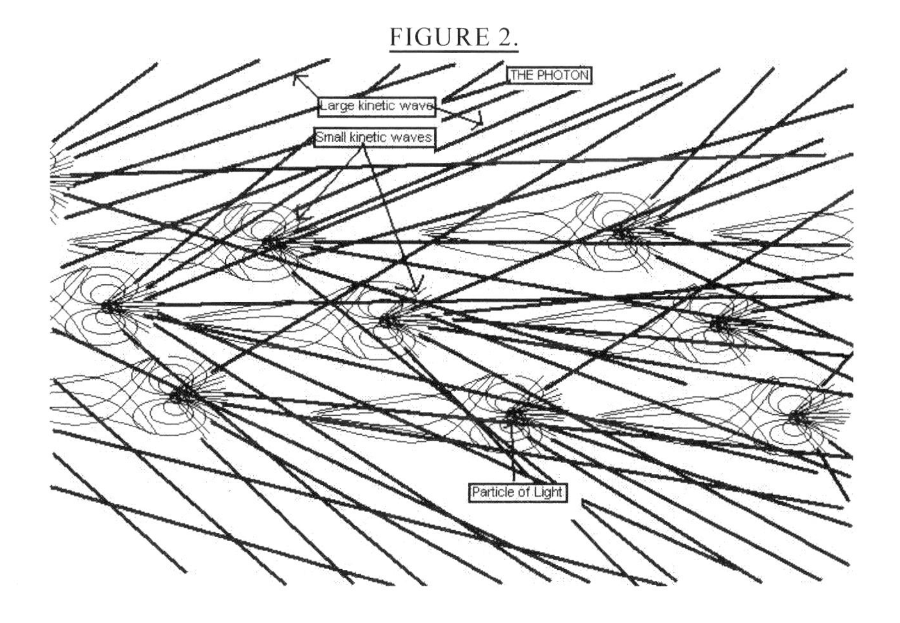

It must be noted that, in relation to Black Body emissions, when the magnitude of the heat being applied to the Black Body reaches a certain degree, the frequency of the light being emitted starts to drop off. I believe, this is not because less particles of light are being emitted during this period of time, but because the number of the particles of light within the photons being emitted during this time, have reached the magnitude, to where they start interfering with each other's respective waves and measuring the exact frequency becomes difficult.

This strong kinetic wave I just spoke of, which physically projects ahead of the photon, also solves Young's double slit phenomenon. This intense kinetic wave reaches the double slit before the photon does and sets up an interference wave just beyond the slits. The center portion of the photon's kinetic wave is weakened by the area between the slits and the two portions that go through the slits then spread out and interact on the back side of the slits. This is why the photons are deflected into various directions after passing through the slits. (See Figure 3.)

FIGURE 3.

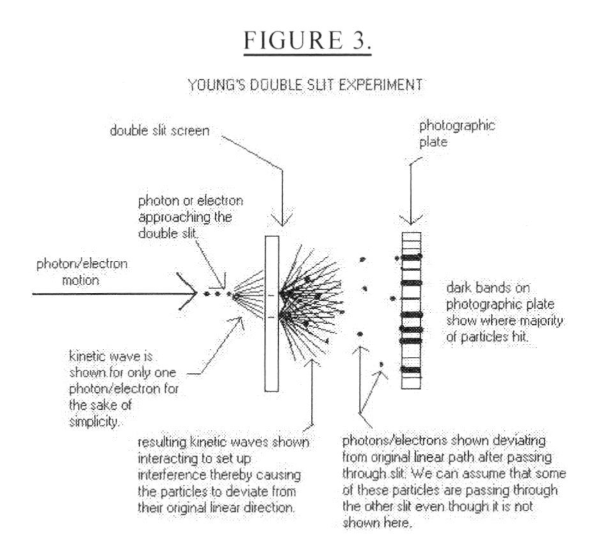

When this experiment is carried out with just one slit open, there is basically no effect on the photon as it goes through the slit. The kinetic wave is there but it is not an interfering wave, as it has no sister wave with which to interact.

A possible way to verify this postulate would be to set another photon gun to the side of the double slit screen and another photographic plate on the other side of the double slit screen. Then simply shoot photons just behind the screen, through the area of the postulated kinetic wave, just before the photons shot from the original gun arrive and record the results. Even if the guns fire at random, eventually the data should confirm one way or another whether this kinetic wave exists or not. I don't believe you will be disappointed (See Figure 4).

FIGURE 4.

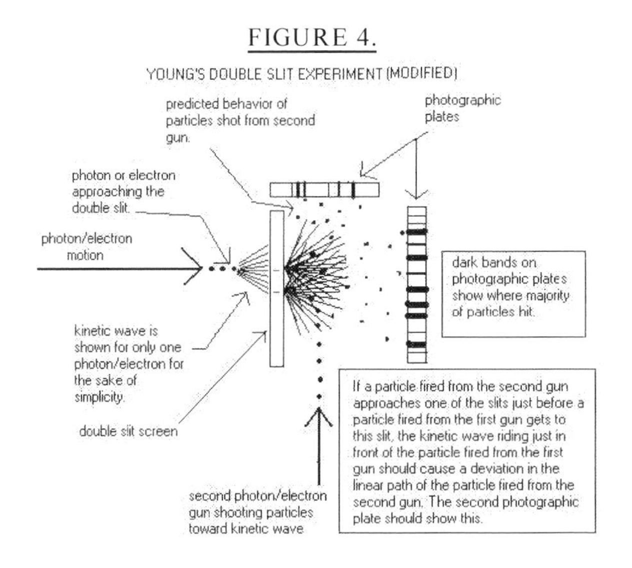

Some might wonder why this kinetic wave, if it exists, hasn't been discovered yet. There was and still is, a problem relating to the effect now known as "self-energy", within the realm of high-energy physics that had to be accounted for by the theorists, in their quest to unify the forces. The self-energy problem is presently seen as the effect a particle's internal forces apply to itself, when in motion. I see this as direct evidence for showing that a particle can interact with its own matter waves.

Einstein speculated the existence of maybe some kind of Ghost Wave relating to this double slit phenomenon. But we must remember that at this point in time he believed that space was nothing or superfluous. He was looking right at the prize and didn't recognize it. What I've shown here can be deemed, by some, as unworkable but take notice that, at the least, it does offer a possible explanation for a great mystery. That's more than any other hypothesis has done so far.

As for the equivalency of inertial and gravitational mass, put me on a space ship and let me guess whether I'm being accelerated or sitting still in a gravitational field. I will simply pull an electron from my pocket and if it's emitting light, above normal, I'm being accelerated, if not, I'm setting still within a gravitational field.

Let us touch again on the acceleration of light problem. This is a phenomenon that says, when a photon is "created" and emitted from any mass object, it accelerates from zero to c, without going any speed in between. In other words the photon accelerates from zero to 186k miles/sec. without running 10k miles/sec., or 20k miles/sec., or even 1 mile per hour. This situation stems from the belief that photons are produced by objects of mass, whereas my model shows that mass objects (quarks) are actually created by orbiting particles of light. And because of this I do not encounter this "instantaneous light acceleration" phenomenon. In my model light does not have to accelerate, but is already moving at c within all quarks which make up all mass objects. As was noted previously, the Big Bang's output was kinetic energy (light) which then went on to create the quarks, which goes on to create all objects of mass. Example, protons contain 3 heavier quarks, electrons contain 1 intermediate quark and neutrinos contain 1 light quark. With that said; all particles/objects of mass that were created as a result of the "Big Bang" can be reduced to a basic particle moving in a very small orbit at c. (This "orbital" speed can vary depending on the object's motion through space). We have come to know this basic particle as light. Light particles were created when the fabric making up space was subjected to pressure so extreme it caused forced ejection of these particles (Big Bang). You could also say that the material space consists of, ruptured under pressure, causing the ejection of some of the material the fabric of space consists of. This caused a huge chain reaction within space itself. The resulting output was a huge, but not infinite, amount of radiation (light). As an aside: this configuration would necessarily allow for large planets similar to Jupiter forming planetary systems around themselves. We only have to look for them.

I need, at this particular point, to introduce the fact that four basic things are happening at once. Space, under extreme pressure, ruptures (Big Bang). A huge amount of kinetic energy (light) is ejected from said rupture. With this huge amount of ejected kinetic energy begins the generation of gravitational fields in space itself. In addition, with the ejected kinetic energy comes the beginning of what we call entropy. We can assume that within the area where space ruptured before the Big Bang entropy = nearly zero. Pressure was the lone force present at this point in "time". I can say "time" here because "time" in fact, is nothing more than a man-made idea used to measure motion during actions. I have found that motion is the base factor for what we call time and I have no doubt that undulations of pressure waves were constantly present and active within space even before the Big Bang. These pressure waves (gravity waves) within space will some day, I believe, be detected in connection with Novas.

I need to make it clear that light (kinetic energy) is not truly created but only transformed back and forth between making up what we call space (dormant energy) and then, after transformation (Big Bang), what we call kinetic energy (light). Individual light particles, without momentum and undergoing the Lorentz contraction in reverse, are actually the particles used to make up what we call space. As all this space structure was expelled from the "Big Bang" through a super heated and super expanded space, at an initial speed above c, the situation occurs where the leading kinetic space matter is slowed as it meets the cooler denser space. This slowing in the speed of the front edge of the ejected space matter (radiation) causes a piling effect when the trailing matter, which is still moving above c, starts over-running it. Another event that is taking place at this same time is that; as all this substance is ejected, the particles making it up, once subjected to this movement, undergoes the transformation known as the Lorentz contraction, which causes them to contract. This is the event in which the condition is set up for mass particle creation. I know I'm being redundant but I ask for patience.

The first logical step would be quark generation (a group of light particles orbiting with one another). Light, without any outside restrictions will travel in a straight line but due to the piling effect we just discussed they are forced into chaotic lines of travel. This chaotic movement, which starts the space around these particles of light spinning, plus the individual gravitational fields associated with each particle of light, eventually turns into orbits, but only around each other, and this builds a particle that has no nucleus but contains a huge amount of kinetic energy compared to its size. Mutual attraction (gravity) within these quarks, forces these particles to interact with each other. The fact that all these individual particles of space (kinetic energy), which were ejected from the Big Bang and became quarks, have the same basic characteristics such as size, weight and speed is why they fail to form a nucleus within their orbits. Again; within this work any elementary particle of mass having no nucleus is in fact a quark. This would include, at least, quarks, electrons and neutrinos.

The particles of light within the quarks are the reason for all other particles having the mass they have and is the direct cause for the property we call gravity. As the particles of light move in their orbits, within the quarks they make up, they generate a gyroscopic effect I call gyre/inertia. Science has long known that matter has inertia and mass but until now, never knew just what mechanical action in nature caused it. The quark basically has a matter wave that is generated as the particles of light within, move through their orbits and this sets up a pulse wave, which spreads in 360^0. And this pulsing (frequency) wave consists of driven space, which has a direction perpendicular to the wave motion. This is the mechanical action that gives the quark its mass. Scientist have long wanted to change the term "spin" to one of 'gyre". Now they can understand why both terms are at play at the quantum level. The light traveling thru their orbits within the quarks generates inertia/gyre, the quarks themselves can revolve which produces spin and the space surrounding the quarks reacting to the particles of light in motion within the quarks produce what I say is mass (matter waves or the quantum force). I explain mass elsewhere in this document and discount the idea that it is a particle. I found mass to be a space based phenomenon generated by kinetic energy within quarks.

Quark creation is absolutely a symmetry governed event. This stems from there being a finite number of light particles ejected from the Big Bang. The heavier quarks formed first after the Big Bang occurred, and the ones that were formed early in this period of particle creation had plenty of the particles of light with which to form themselves. These early heavier quarks (top) became the most massive because they had particles of light filling most or all of the possible orbits within them. I use an extension of Pauli's exclusion principle for this particular orbital configuration within the quarks. But what, we now must ask, is the "binding" property at play here?

It should be noted here that gravity, within this model, is in fact, not an active force at all. The gravitational field is generated not by mass but by the individual particles of light within and around mass. Show me light and I will show you the entity responsible for gravity. Show me light and I will show you a gravitational field. Light expands space to generate an inertial guiding path of least resistance. The light particles making up the quarks, from which all other particles are made, is the single entity most responsible for what we call the gravitational field. It has been commonly assumed that the earth, sun and all other objects containing mass radiate a particle (graviton) which generates an attractive force called gravity. This monopole particle has yet to be found and the reason why should become obvious to anyone who reads this work, now that I have mechanically explained just how the effect of gravity is created. In connection to the above; scientists will readily admit that they have absolutely no physical/mechanical idea for what the property called "charge" really is. Charge has been extremely useful in that it allows for a possible force needed to hold atoms together. Under my model charge is unnecessary (superfluous) for gravity and matter waves (quantum force) work just fine for creating atoms of every type.

Ask yourselves, with reference to the Big Bang theory, what was the initial output from the Big Bang? The answer is a huge plasmic wave of "kinetic energy". Now ask yourselves what was the first presumed force to come onstage after the Big Bang? The answer is "gravity". Can you see the connection here between kinetic energy and gravity? Is it possible that radiation and gravity showed up at the same time right after the Big Bang occurred and there exists no connection between them? It can, I believe, be truthfully stated that; gravity is the result of space reacting to kinetic energy traveling thru it.

I find here that I must amend the Big Bang Theory thusly; the one and only force is pressure. It causes the Big Bangs, it then drives the individual particles of light that came out of the Big Bangs, which then goes on to expand space and generate the gravitational fields, which causes particles of light to be trapped into orbits about themselves which produces quarks and later forces all the bodies of mass (quarks) together in order to create protons, neutrons, electrons, neutrinos, atoms, molecules, planets, stars, and galaxies. This is the ultimate Occam's razor. Can it all be just this simple? I believe it can.

CONCERNING THE SPEED OF LIGHT POSTULATE

This work will attempt to point out a problem arising from the postulate stating, "The Speed of Light is independent of both the motion of its source and also the motion of any observer". Albert Einstein, on pg. 17 in his book, Relativity- The Special and the General Theory, referenced the work of a Dutch astronomer De Sitter as the source for the first part of his postulate, which states, "The velocity of the speed of light cannot depend on the velocity of motion of the body emitting the light". De Sitter arrived at this conclusion following observations he made relating to double star systems. And I "absolutely" agree with his findings.

I will not address this part of his postulate within this work, but will address the second part of the postulate, which states, "The Speed of Light is also independent of the motion of any observer". Mr. Einstein ties this part of the postulate to the works of Maxwell, H. A. Lorentz (pg.40), and also to the Michelson/Morley experiment (pg. 53), performed in order to prove the existence of what was once called, "The aether", or better known as, "Newton's rigid and non-moving space". The experiment failed to exact the results that were sought. The negative result however, led Mr. Einstein to a separate conclusion. He reasoned that since the earth was in motion at some respectable velocity around the sun, and observers on earth still measured the speed of light to be 186,000 mps, this pointed to the independence of the speed of light, to the motion of these same observers.

I will not challenge this reasoning within this work, but will allow it to stand along with the postulate that evolved, as a result of this same reasoning. My intention is to now show the problem that arises, when using this second part of the postulate in calculating distances traveled by objects, through space and over time.

My initial attempts were met with the same patented response, which was, I did not understand all the relativistic effects that existed within my models. As time moved on I came to realize that as long as I inserted different inertial frames of reference into my work, I would always have this same problem.

It was not that my approach was wrong, but more the complexity, I needlessly injected into my setup. So I devised a much simpler approach to get around this problem. I made all distance measurements for the photon, in my next setup, from just one inertial frame of reference. I did away with all stationary reference frames and inserted two spaceships along with a single photon. The initial setup now has one spaceship traveling in a straight line through space at 99% the Speed of Light. The second spaceship passes just in front of the first spaceship and is traveling in a straight line perpendicular to the direction of the first spaceship. This second spaceship is also traveling at 99% the speed of light. This effectively causes both spaceships to experience the same relativistic effects. I now have a single, inertial frame of reference, with which to make my measurements, which will make the measurements taken by either spaceship equally valid when compared to each other. Using this approach I only need to have the photon pass between the spaceships, and traveling in the same direction as the second spaceship as it draws even with the first spaceship. This is the point in this setup where the measurements/calculations are to be made.

Using the Laws of Physics and the postulate, "The Speed of Light is independent of the motion of any observer", we now calculate where the photon will be 1 second from this point in time. The distance measurements for the spaceships and photon will be calculated by observers on both spaceships. Be aware; the clocks on both spaceships have the same relativistic effects being applied to them and this 1-second time period will be equal for both spaceships. This means that different relativistic times can now be discounted within this setup and the experiment can now begin.

First we find that the observers on both spaceships each measured the photon's speed to equal 186,000 mps as they passed each other and as the photon passed between them. This is in agreement with the postulate. Now the individual observers on each spaceship calculate where the other spaceship and the photon will be, 1 second into the future.

This is where the problem appears for the part of the postulate that states, "The Speed of Light is independent of the motion of any observer". Using this postulate gives us two different values relating to the distance the photon will travel in this one second.

The first spaceship, which is traveling at a right angle to the direction of the second spaceship and also the photon, calculates measurements that say, itself and the second spaceship will travel 99% of 186,000 mi. (184,140 mi.) in this one second, and the photon will travel 186,000 mi. during this same second. The second spaceship calculates that, itself and the first spaceship will travel 99% of 186,000 mi. (184,140 mi.) during this one second, which is in agreement with the calculation made by the first spaceship. But the second spaceship calculates that the photon will travel 199% of 186,000 mi. (370,140 mi.) during this same one second. This is due to the postulate forcing the second spaceship to calculate, that the photon will be outrunning it by 186,000 mps during this one second. So the distance that will be traveled by the second spaceship during this 1 second, must be added to the 186,000 miles calculated for the photon, which will be advancing ahead of the second spaceship at 186,000 mps, during this one second (See Figure 5.)

FIGURE 5.

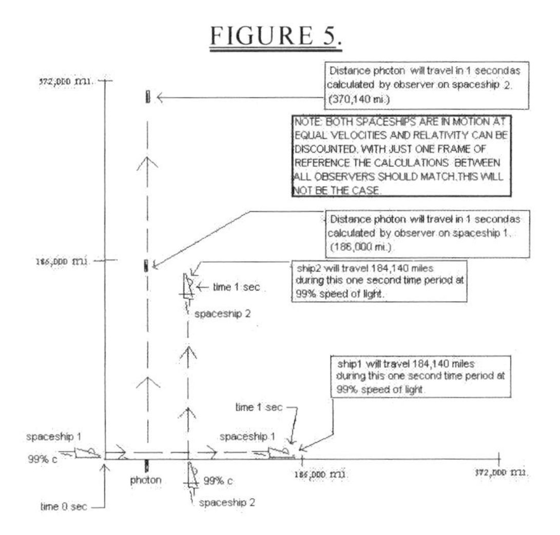

We now have the same photon calculated to be in two different places at the same time, and the difference is far beyond what any minor correction to this postulate can account for. As I see it, we have only two choices in reconciling this difference between measurements. Either the direction through space, can affect the distance a photon will travel during this one second or the postulate is in error. I leave it to you to make your own assessment.

The event I just described was a hypothetical event, but all the conditions and requirements I imposed were within realistic boundaries. By this I mean, at some point in the future, this hypothetical experiment will be possible to perform, given our present rate of technological advancement.

I will now give you another situation, which points to the same conclusion as the situation above. Imagine being aboard a spaceship that is setting stationary in space (in a space dock) and a photon passes by it. An observer on the spaceship (reference the dock) records the speed with which the photon passes it and the result comes out equal to 186,000 mps. And now the photon has passed the spaceship and is receding from it at this same speed. After a moment the observer decides that he wants to chase the photon, so the spaceship fires up its engine and accelerates to ½ the speed of light. The observer now measures the rate, at which the photon is advancing from the spaceship, as it chases this same photon at ½ the speed of light. According to the postulate, this measurement must come out equal to the measurement taken while the spaceship was setting still. This means that the rate of recession between the photon and the spaceship will always be 186,000 mps, whether the spaceship chases it or not and requires a huge time adjustment. Too go even farther, we now turn the spaceship around and accelerate directly away from the photon at ½ the speed of light and we now measure the recession rate. Again, according to the postulate, we will measure the rate of recession to be equal to the rate of advancement (186,000 mps). At this point we need to do an in-depth analysis.

The postulate, when broken down to a "logical" base definition says, "In reference to the photon, the motion of any observer will always be zero". We back this by the example above where the speed and direction of the spaceship had no effect upon the advancement or recession rates between it and the photon. In other words, the photon sees itself as the only object in motion or independent from motion of any other source. This is the meaning of the postulate in its rawest form. And using this definition, we now emit a photon towards the spaceship from 1 light year away. Under the situation we have just created and employing the postulate, with reference to the photon, we must find that, even if the spaceship powers up and accelerates to 1/2 the speed of light in a direction towards the approaching photon, it will still take the photon a whole year to reach the spaceship. Remember, with reference to the photon, the spaceship is not moving. The question becomes; which part of reality do we discount with reference to this event?

As a side note, if we substitute the word "object" for "observer" within the postulate, and I find no reason for this not being allowed, we can now say, "The Speed of Light is independent of the motion of any object". The object can be source, observer, or non-observer and it will make no difference. But if the object happens to be another photon then we have a case where the postulate runs over itself again, for the other photon claims the same laws of motion as the first photon. If the "Constancy of the Speed of Light Law" is valid for the first photon then it must be denied for the second photon, with reference to the first photon. I find it more than a little strange when a theory, taken to its ultimate conclusion, denies its own claims. Under this postulate only one photon could be in motion at any one time within the universe as a whole. I cannot make it any clearer than this, in showing why there exists a problem with the postulate.

In retrospect, the cause for this confusion can be laid at the door of one event that occurred in the past. That event is the experiment performed by Michelson and Morley in their effort to prove the existence of the aether (physical space). Michelson violated the most basic rule in experimental physics when he inserted a device into his setup that was neither defined nor understood. That device is what we call light. The scientific community has yet to prove the cause for the speed with which light travels. Light is also thought to be paradoxical in that, it can appear as either a particle or a wave. The bottom line is that this particular experiment, because of the unknown element within it, should have been rejected along with its results. I assert that, this one flawed experiment has put science on a sidetrack for over one hundred years.

CONCERNING THE SECOND PART OF THE SPEED OF LIGHT POSTULATE

The Speed of Light Law states, "The speed with which light travels (in vaccuo) is independent of both, the motion of its source and the motion of any observer". Put simply, "Light travels thru "assumed" empty space at one constant velocity". But we must also realize that all other matter travels thru this same space with either constant or varying velocity depending upon frames of reference. With this said, let us add an observer to the equation.

First we make our sun and the observer's eyes the only frames of reference. We then place the observer out in space away from the sun. The observer will see the sun as a mostly white glowing orb. Now let us subject the observer to an extreme acceleration in a direction away from the sun. The observer will now see a shift in the sun's color toward the infrared portion of the spectrum of light. Under this circumstance we need neither a clock nor a frequency analyzer to back up our finding. The observer knows without a doubt that the sun's light reaching his eyes, while he is in this motion, is somehow different. But what is causing the difference? There can be only one answer; the observer's motion!

The suns energy (light) is hitting the observer's eyes with less intensity when he is in motion away from the sun. This is due to the fact that the rate of approach between the sun's light and the observer has been decreased. Note: The sun does not emit less light simply because an observer puts himself in motion. It's just that it will take longer for light waves to pass our observer in motion than it will if the observer is sitting still. Hence, the shift!

In conclusion and with reference to the observer's eyes: The observer will deduce very clearly that the sunlight's rate of approach, is greater when he is stationary, than it is while he is in motion. And all this while light was still traveling thru space at a constant velocity! CLOCKS BE DAMNED!!!

THE DOPPLER EFFECT

The Doppler effect is used to show wave characteristics relating to air, water and light motion. Science at present has no problem with understanding air and water motion but a phenomenon arises when it comes to addressing light motion. The problem is that air and water are both recognized as mediums but since space is presently seen as empty it cannot be used as a medium for the light traveling through it. But with this approach, the question now becomes; "What then is waving"? In my judgment there exists only one potential answer; space itself.

We will now do a comparison between these three wave entities. If we hit a bell with a hammer, the bell and hammer stay in place while the surrounding air absorbs the sound (air waves) that was released due to the bell being hit. We might see this as a reaction, by the molecules of air, to the shock being applied to the molecules within the bell by the hammer. The "shocked" bell is the action and the resultant response of the air around this "shocked" bell (sound wave) is the reaction. This same process applies to the act of dropping a pebble into a pond only the reaction is seen as a water wave. Note that the resultant air wave spreads out in a spherical 360^0 configuration and the water wave spreads out in a two dimensional 360^0 configuration. The main point to remember here is, one hit applied to the bell produces one wave set and one pebble dropped into the water produces one wave set. Fourier analysis describes these wave functions very well and this model agrees with what it shows.

So now let us hypothetically, take one particle of light and put it in motion through space. This would be a photon with a frequency of one. The energy content of this photon with a frequency of one would turn out to be equal to 6.626×10^{-34} joule second or better known as Planck's constant. The energy value (Planck's constant) is not the energy connected with the particle of light but is the energy value of the kinetic wave that projects ahead of the particle of light.

Within this model, the wave/particle duality of light is not a phenomenon, but is a real, physical part of nature around us. This means, if we dissect a photon with a frequency of one, we will find one particle, surrounded by one frontal wave and a responsive transverse wave behind. Einstein said that only by treating light as if its wave/particle duality was real, can we hope to understand it. I agree with this assessment with the exception that I exclude the "as if" and treat the wave/particle duality as a true, physical part of nature. My analysis shows a particle of light in motion through the medium of space (action), causes space to generate a wave set around this particle of light (reaction). This is the only way that the wave/particle duality of light can be experimentally shown to exist as a legitimate part of nature around us.

And by use of the wave/particle duality, we see a difference between the configuration of the air and water waves when compared to the light wave. With air and water the action mechanisms (hammer and pebble) stay in place and the resultant waves are nothing more than molecular interactions through the mediums, by the mediums. But with light we find that when a quark (electron) is caused to emit a particle of light, the particle of light accompanies and continues to generate the resultant spatial wave that it generated.

Let us now look at the Doppler effect relating to light in motion, only with these new parameters applied. First we assume that the speed of light is not independent of the motion of any observer and secondly we assume that space is a true medium for the light traveling through it. In addition we also make the area in space where the Big Bang occurred the inertial reference frame **(^)**. Now let us look at how this new view works in reality.

First off, we must assume that the earth is moving through space at some speed and in a direction generally away from the area in space where the Big Bang occurred. Let us assume that the earth is traveling at approx. 40% c away from the area where the Big Bang occurred. We can discount the earth's orbit motion for this situation. If we take a photon with a frequency of two and transmit it towards a spaceship, in motion through space at ½ c in a direction away from us, the photon will approach the spaceship at ½ c, with reference to space (^). This means the two particles of light within the photon, will arrive at the spaceship with a slightly larger distance between them (red shift) than they would if the space ship was in motion at the same speed as the earth. We get the opposite result (blue shift) if the spaceship is in motion towards us.

The point to be made here is that the particles of light and their transverse waves travel through space together and the shifts we see can lead us to only one conclusion. If the speed of light was truly independent of the motion of any observer then no shifts could occur. Only by stretching out or compressing the distance between the waves/particles do we get a shift and the only way to stretch or compress the waves/particles is by the relative motion of the observer (spaceship). Let us now expound upon the speed of light postulate to a greater degree.

In (Figure 6.) I present a hypothetical work showing the postulate being applied in a basic way (See Figure 6.). The observers are positioned on the spaceships a and b. Observer A is on ship a and observer B is on ship b.

FIGURE 6.

1. Given: The speed of light is independent of the motion of any observer.
2. Two spaceships moving through space at .99 (c) on parallel paths.
3. A photon passes between them on the same path.
4. As the photon passes between, the observers on both ships calculate where both the ships and the photon will be, a second later.
5. With both ships in motion at the same speed then, according to relativety, their clocks will be the same since they share a mutual reference frame.
6. Both observers calculate the following: In 1 second both ships will travel 184,140 miles and because the photon, according to the postulate given in 1., has been out running them by 186,000 mps, then the photon will have traveled 370,140 mile in this same second.

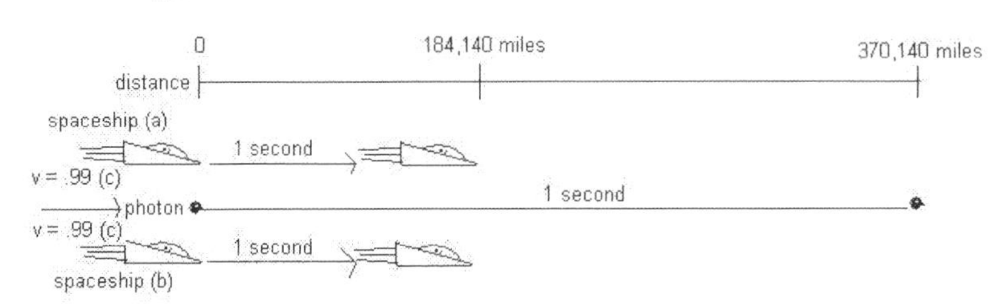

As we discussed earlier in this work, if the speed of light is truly independent of the motion of any observer (frequency counter), then how can the frequency shifts (wave expansion/contraction) occur"? How can one retard or increase, with ones own motion, the distance between the light waves (frequency), when the speed of these same waves are supposedly, independent of your motion? The choice, as I see it, is between Dr. Einstein and Doppler. I choose Doppler.

I have no doubt that Dr. Einstein was thinking in terms of the time recorded by observer O2 and his clock C2 who were in a motion away from the source L of the light wave. If this be the case, his postulate needs to be amended to allow for the differences between the clock of the observer in motion and the clock of the observer positioned at the source. Now it is time for some calculations (See Figure 13).

If we changed the single light wave emitted from source L to a photon with a frequency of 10,000, then observer O1 at S, which is at rest relative to source L, would have experienced 10,000 waves passing him every second, hence he would see no shift. But observer O2, who was in motion ahead of the light wave at c/2, would have experienced, with reference to frame (S), 5,000 waves passing him every second. But we must also take into account that observer O2's clock is slow when compared to observer O1 at frame (S). Hence, observer O1 would record, less red shift (more than 5,000 light waves/sec). But how much more?

On page 87 in the book RELATIVITY-The Special and the General Theory, Dr. Einstein performs a time and space calculation using the first and fourth equations of the Lorentz transformation. In relation to the measurements for time (1 second), with reference to a body in motion, he comes up with an equation that says, the difference in time-keeping, between a clock at rest and a clock in motion, can be stated mathematically as: $1/\sqrt{(1-v^2/c^2)}$. For the clock C2 in motion at c/2, this comes out to be 1.1547005 seconds, with reference to the clocks C1 and C3 at rest and their 1 second. This gives us a difference of .1547005 seconds.

The clock C2's period of time for 1 second, is longer than the period of time for 1 second, with reference to the clocks C1 and C3 at rest, and in this case, .1547005 seconds longer. But note that in reality, the actual frequency of light, with reference to its source L and its speed through the space around it, does not change. Or we can say, "in reference to its source L, light and its wave passes through the space around it, with no change in its frequency. With reference to observers O1 and O3, observer O2 who is in motion at c/2, is experiencing a frequency of 5,000 waves/sec. But note that this frequency (waves/sec) is with respect to clocks C1 and C3.

When we plug clock C2's extra time into the situation above, this means more light waves than 5,000, will pass observer O2 and his clock C2, during his 1 second than will pass observer O1 and his clock C1.So let us now add this extra time and recalculate.

We know that light with a frequency of 10,000 is passing observer O1 at frame S each second, relative to O1's clock C1 and also relative to clock C3 at source L. We also know that light with only a frequency of 5,000, in reference to clock C3 at source L, is passing observer O2, because of observer O2's motion at c/2. So now we add the additional time .1547005 seconds. This calculates out to be a frequency of 5773 with reference to observer O2 and his clock C2. So what does this say?

With reference to the source L of the photons, with a frequency of 10,000, an observer O2's clock C2 in motion at c/2, in a direction through space away from the source L, will record a red shift, or reduction in original frequency, of 4,227 waves per second.

Now let us reverse observers O2's direction, and show him in motion towards the source L at speed c/2. Observers O1 and O3 with their respective clocks will see 15,000 waves/second, passing observer O2 in this configuration. Observer O2 with reference to his clock C2, which is in motion at c/2 and running slow by .1547005 seconds, will record 17,321 waves pass him during one second.

With reference to the source L of the photons, with a frequency of 10,000, an observer O2 in motion at c/2, in a direction through space toward the source L, will record a blue shift, or increase in the original frequency, of 7,321 waves per second. This explains why an observer's motion towards the source, results in a larger blueshift than the redshift that occurs, when an observer is in the same motion away from the source. In addition, note that only with the frame of reference **(^)** in play, do we have a chance to fully understand how relativity and the Doppler effect, really works in nature around us.

I will now show you an interesting fact. In the work above I used the imaginary frame **(^)** and also two different and relative approach rates, one with the observer O2 in motion away from the light waves and the other with the observer O2 in motion towards the light waves. Understand, since we can't outrun or even equal the speed of light, this means that you will always have an approach rate between you and the light waves, whether you are in motion away or towards the light. Dr. Einstein, according to his principles within the law for the constancy of the speed of light, believed this approach rate would always be equal to the speed of light, no matter the motion an observer might be under, but I will show this to be untrue.

With observer O2 traveling at 93,000 mps ahead of the light waves, which have a source frequency of 10,000 and after factoring in his clock C2's dilated time, he records a frequency of only 5773 (red-shift). The approach rate I calculate here, between observer O2 and the light waves, was 93,000 mps (c/2). But Dr. Einstein's law says, light will approach any observer, no matter his motion, at only speed c.

With observer O2 traveling at 93,000 mps toward the light waves, which have a source frequency of 10,000 and after factoring in his clock C2's dilated time, he records a frequency of 17, 321 (blue-shift). The approach rate here, between observer O2 and the light waves, was 279,000 mps (1.5c). Again, Dr. Einstein says, by way of his speed of light laws, light will approach any observer, no matter his motion, at only speed c. It would seem, according to this law, I have a problem. So if I am wrong, according to Dr. Einstein's laws, let us plug our data into the Doppler equation and see just how wrong I am.

As I analyze the Doppler equation and its rules, I find that this equation uses variable approach rates, the same as I do, but hides them in the math. In particular, velocity must be given as a negative value to make the calculations work, when the observer is in a motion away from the oncoming light waves. As we well know, velocity cannot have a negative motion, only positive. Equations should reflect reality whenever possible.

In addition, I see that someone has substituted 1 for the speed of light c and does it twice (1 + v/c) over (1 – v/c). If we look up the laws of division it says, the denominator sets the units and the numerator gives the number of these units. So what is the denominatorial unit when the observer is in motion at c/2, toward the light waves? It is 93,000 mps (c/2). And what is the denominatorial unit when the observer is in motion at c/2, away from the light waves? Its 279,000 mps (1.5c). These values, mathematically, can only reflect one possible relationship within the Doppler equation, and that is rates of approach between the observer and the light waves coming at him.

I believe I have simplified the equation and at the same time, made it easier to understand. (See Figure 8.)

FIGURE 8.

DOPPLER EQUATIONS FOR LIGHT

Original Doppler Equation for light

$$f = f_S \sqrt{\frac{1 + v/c}{1 - v/c}}$$

f is the observed frequency.

f_S is the frequency of the source.

v is the relative speed between the source and observer.

If source and observer are approaching each other, v is seen as +.

If source and observer are receding from each other, v is seen as –.

Equivalent Doppler Equation for light

$$f = f_S \sqrt{\frac{c + v}{c - v}}$$

f is the observed frequency.

f_S is the frequency of the source.

v is the relative speed between source and observer.

$\sqrt{\ }$ incorporates the relativistic time factor.

The terms, (c + v) and (c - v) are (rates of approach) between the observer and the approaching light waves.

If observer is advancing toward the approaching light, (c + v) is the numerator.

If observer is advancing ahead of the approaching light, (c - v) is the numerator.

In conclusion, I want to make it clear that this work, in no way, challenges the Law for the Constancy of the Speed of Light in vaccuo or the principle that says, the speed of light is independent of the motion of the source. (Remember that within this work, light (energy) is always in motion at a constant velocity, no matter its location). But this work does challenge the principle that says, the speed of light is independent of the motion of any observer. It also challenges the idea that space itself, is empty (non-physical in makeup).

In addition, it must also be understood, with reference to a physical space, all measurements involving time with reference to earth, are "relatively" dilated (slow) in respect to their accuracy. I believe that our measures of time can be altered by motion through space, but whether Lorentz's equations, relating to these time alterations, are in fact correct, remains to be discovered. In addition, the personnel who work with Quantum Mechanics must recognize that everything in nature is truly "quantized" to include all forces (and gravity) and even space itself.

LIGHT (GOD'S PUTTY)

To begin, we must recognize that all elementary particles of mass matter are composed of kinetic energy (light traveling in orbits about each other) and we call these elementary particles quarks. When I say kinetic energy I mean that the energy (light) within the quarks is in motion at speed c. As I have stated elsewhere; all elementary mass particles are, in reality, quarks. Quarks in a triplet configuration form the protons, single quarks make up the electrons and neutrinos. Note: Any elementary particle having no nucleus is, in fact, a quark. Quarks supply all particles with their mass property. But what happens when the universe runs down and finally the quarks start to decay?

First we must realize that the photon is a particle that man has devised in response to his measurements relating to samples of light waves going through detectors. The photon is simply a description of particles of light, moving in a linear direction with a certain frequency. Within this model, the photon is the state all mass matter must eventually reassume. This model also dictates that the decay or partial decay of a quark, creates photons. The difference between partial decay and full decay is that partial decay is connected with light frequencies in the infrared and visible range, while full decay is connected with the ultraviolet, and, gamma rays can be emitted when numerous quarks decay. The quark has no nucleus and is made up of light particles locked in orbits within it, by their mutual gravitation fields. The quark was the first particle to be created after the Big Bang occurred, with the exception of the particles of light that it is made from. It then goes on to make up all the matter in nature around us, with the exception of space.

Max Planck theorized through statistical analysis, that energy emissions from matter have a finite lower limit and all photons have this value as a multiple within their makeup. This value is calculated to be 6.626×10^{-34} J s, and is known as Planck's constant. Within this model there does exist an energy source small enough to be equal to this tiny value called Planck's constant and it is directly related to light.

As I stated before, the particle of light, which is the basic particle making up all of matter, to include even the fabric of space itself, causes space to generate a kinetic wave just ahead of the particle of light as it travels through space (See Figure 9.). What do you guess the energy value for this tiny kinetic wave might be? Its energy value is equal to Planck's constant. This is why this tiny number is inherent to every photon that was ever emitted. I need to make it clear that the energy of this kinetic wave riding just ahead of the particle of light is not the energy of the particle of light but is the reflective kinetic energy (force) of space reacting to the particle moving thru it. The particle of light retains its contracted form within itself and it is the force of pressure applied by space, which gives it its speed. In actuality, space squeezes the particle of light through itself. The result of this action is a kinetic wave forming in space just ahead of the particle of light and this kinetic wave becomes the force that generates all the other forces in nature around us. And the energy connected to this kinetic force wave equals Planck's constant. See Figure 9. Note also: these single Planck particles of energy is all the gravitational fields see as real and interact with.

FIGURE 9.

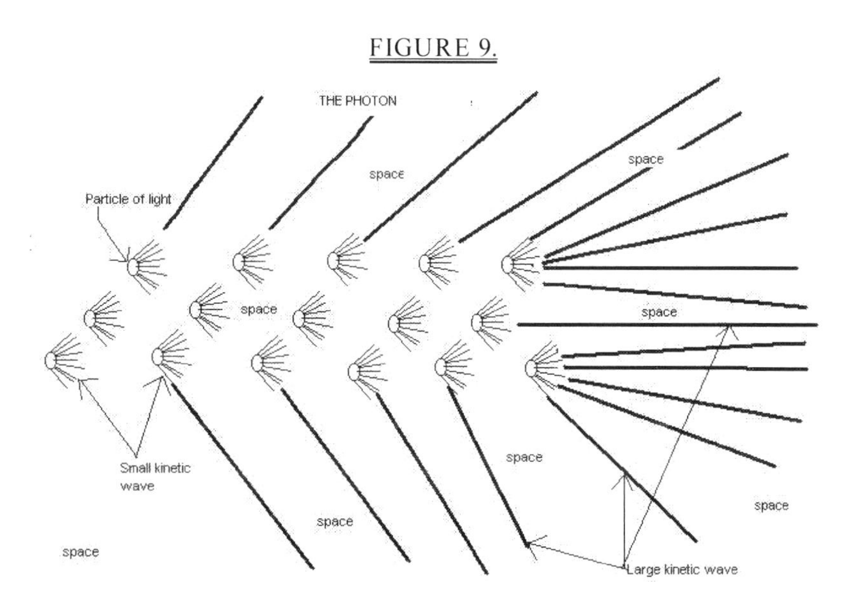

Newton argued that light must be particle based. Huygens argued that light must be wave based. Since then, this debate has raged over several hundred years with plenty of experimental evidence being presented to substantiate both sides. This in itself should have alerted the scientific community to the possibility of mutual collaboration between the two phenomena. From what I have shown it should become clear that the universe itself exists as a particle/wave duality. Even the particles making up the fabric of space can be considered as a product of particle/wave duality. Mass matter is a result of kinetic energy (quarks), space is dormant energy.

Something that immediately falls out of this postulate is that I can now tell you how many particles of light are in a photon simply by knowing its frequency. Let us take a photon in the yellow-green range of light, having a frequency of say 540×10^{12} Hz, and apply Planck's formula, which says the total energy of a photon can be measured by multiplying its frequency by a constant, whose value is 6.626×10^{-34} J s. After the math we end up with 3.578×10^{-18} J s for the energy of this photon.

So basically we can say that the frequency of a photon is equal to the number of light particles in that photon. An exception to this would be when great heat is applied to an object, which would cause the particles of light being emitted, to become entangled with each other's transverse waves, and under this scenario, the frequency would not correctly reflect, a proportional amount of energy with reference to a frequency counter.

The one constant in any interaction within and between forms of matter is light. From the spectacular action of high-energy collisions all the way down to the subtle nudging of electrons, there is one ever-present side effect, which is light. The mind-set of the scientific community at present is that matter creates light but I will now present an alternative to this view, which will also explain the phenomenon connected with the present view.

The present belief, concerning photon creation, leads to a phenomenon relating to the instantaneous acceleration of light. This phenomenon does not exist within my model for it demands that light make up matter, rather than the other way around, and when this light shows itself, in the form of photons, it is already traveling at 186k miles/sec. Think about it.

Light is emitted in distinct groups that make up what we have come to call photons. As I said before, what makes the photon unique is that the light particles within it do not move in orbits but form up together and move away from the sources that emitted them, in a linear direction. Like a spread-out herd of horses all going in the same direction or just a bunch of individual light particles, and their individual kinetic waves sweeping linearly through space creating a larger kinetic wave. Because of this characteristic, my configuration of the photon disallows the possibility of an anti-photon. In addition the photon has no inertia/mass. This is why this particle can approach an object at c, and is deflected away in a straight angle. If the photon had mass, when it came into contact with any other form of mass, the result would be catastrophic. Note though, photons, due to their motion, must possess a value of momentum.

In many instances the photon is emitted by a quark that has fully or partially decayed. The quanta values that differ between photons would be governed by the number of light particles lost by the quark in its decay. Quarks can have different configurations, depending on the period of the Big Bang event they were created in. Just after the Big Bang occurred, there was an abundance of light particles available and the heaviest quarks were created then, but after a time, light particles became less available and the intermediate (electrons) and weak quarks (neutrinos) were created during this period.

As time continues, entropy (space pressure) wears the quarks down to just individual particles of light, which are then reabsorbed by the fabric making up space and they lose their momentum. Space can absorb individual particles of light, mostly after this light has traveled beyond the gravitational fields associated with all the matter that was ejected from the Big Bang. A particle of light, once it has lost its momentum, re-expands somewhat to become part of the matter that makes up space again. I say somewhat, because full re-expansion is not possible as space is so full of other light particles also in their dormant or what I call rest-state. This particle of light ultimately ends up in an intensely pressured configuration within the fabric of space itself. And, as more and more light particles end up in this same configuration the pressure keeps intensifying (apply kinetic law of gases here, only under extreme pressure). This mechanical act that the particle of light performs when it comes to rest, which is expansion (Lorentz's expansion), is what causes the kinetic pressure needed for another Big Bang out there somewhere in space and so it goes, on and on. What was just now presented points to a conclusion which says, "With reference to a perpetual universe containing a steady stream of Big Bangs, Hoyle's Steady State Theory, though inexactly, becomes mathematically, more than plausible".

With that said I need to make the reader aware of the different physical descriptions I apply to matter within this work. I was forced into this quandary by the requirements laid out by the standard model. The standard model uses a particle based configuration in its attempt to depict the makeup of reality. This caused somewhat of a problem for my theory requires that the universe is a solid entity under extreme pressure with no voids at any point within or without. Showing the universe, under a required perpetual configuration, in order to correctly describe reality, was the only path I was afforded. All other paths failed to allow explanation for a perpetual/dynamic universe such as we live in now. The key became the fact that kinetic energy is directly connected to the space transmitting it. I was thus driven to formulate a physical/mechanical configuration involving an overall pressure generating action within the matter making up the fabric of space itself. Hence, my theory necessarily uses pressure to cause a rupture within the "physical" space surrounding us (Big Bang).

The connecting particles making up the fabric of space have a "mean free path" between them of 0. The Big Bang occurred when the pressure became so great on the fabric making up space that the "mean free path" between the particles making up the fabric of space became less than zero. This caused annihilation to occur between the particles of space itself as space simply could not restrain itself in this situation and the result was a huge outpouring of kinetic energy (contracted particles of space) in the form of radiation (Big Bang). And he brought forth light from the darkness!!

I'm sure some are aware that in the description I gave before, of single Planck particles of light being absorbed by space, I allowed light to come to rest and become dormant (static) as part of the fabric which makes up space itself. The underlying factor that ties all this together is my confidence that space is extremely hard or in fact, a solid, if you will. When I first started my quest to understand the universe (1969) space was mostly seen as a void. In the past, up until the late 19th century, space was assumed to be physical in its nature. Later, the idea of virtual particles in space was introduced in order to balance black hole equations and more recently, dark energy and dark matter was introduced in order to rectify problems with theories calling for increased expansion rates relating to the universe as a whole. Note the following; "Cosmologists claim, if the inflationary Big Bang model is correct, then the cosmological constant for the expansion of the universe is close to one. In order for this constant to be near one, the total mass of the universe must be more than 100 times the amount of visible mass that appears to be in the universe. This calculation indicates that more than 99 percent of the mass of the universe must arise from dark matter". With that said; presenting my idea of "physical space" was a non-starter when I began this quest but year by year it seems to become more tenable. Time will tell concerning my hypothesis for a possible physical space, need I say more?

Note; angular force (mass) is converted to linear force (pilot waves), as objects are put into motion (De Broglie). Within this model, there is no part of the universe that is ever stationary or truly holding just one position in the space around it, and this includes even the matter making up space itself. The scientific community at present, believe we travel thru space without interacting with this same space. This is because they see space as superfluous (Einstein). I can tell you without a doubt; if space were truly unnecessary then light would have no way to generate the wave connected to it and forces would have no way to interact with each other. This model, by the requirements imposed by the laws of thermodynamics, had to include all these characteristics in order to maintain its infinite and perpetual nature.

CONCERNING MICHELSON'S ATTEMPT TO PROVE THE AETHER

Just exactly what is space? We know from Axiom # 2 that it has to have some physical value in order to exist. The idea that space might be something with a non-moving physical value once reigned as a self-evident truth. This belief was erased by an experiment performed by a man named Michelson late in the 19th century. I have always had a problem with the nature of the way this experiment was performed. Michelson theorized that if space was a physical entity then since the earth was moving through it we ought to be able to detect this movement.

Michelson figured that the earth, moving through this non-moving physical space, ought to be similar to a boat moving across a lake and leaving a wake, and he thought we ought to be able to detect this wake. He then decided to use light as the experimental tool for this experiment. He figured that light projected upstream in this wake and back, would take longer to make the trip than light projected across this wake and back. In principle he had a workable idea but the problem lie in his choice of experimental tools. He violated one of the most basic rules in experimental physics. He chose to use a device within his experiment that was not completely defined or understood. This device is what we call light.

This is how Michelson's experiment was flawed. He did not know that the speed of his measuring device (light) was actually controlled by what he was trying to detect. It must be recognized that the force used to propel the boat through the water must come from the boat itself (self-energy) whereas the force that propels light through space is from the squeezing pressure of space itself. If light truly moved by self-energy, like the boat, he would have seen a difference between the beams he split. But because the speed of light is powered and controlled by space, he arrived at the wrong conclusion, because he did not know this. When Michelson split the beam of light, the two beams of light went off in right angles to each other. Michelson had the idea in his mind that the light going up stream and back, using X amount of force, would necessarily be retarded compared to the light going across the stream and back, using the same X amount of force. He would have been correct in his thinking if light had to propel itself but such is not the case as I have shown.

Within this model, space transmits (squeezes) light at one speed, which is c and since particles of light make up all mass (quarks), this means that if you are on a spaceship traveling at ½ c and a photon passes you, it will do so at ½ c. In Michelson's experiment and in reference to space, the light beam he split came into contact with the splitting mirror at speed c, minus (the velocity of the earth through space). At the same time we must figure in the fact that the reflecting mirrors are in motion and the perpendicular beam has to travel an extra distance out and back. The key here is to remember that earth, minus the orbital motions it makes, like all other mass objects, is moving away from the point in space where the Big Bang occurred at some velocity. I will repeat; this one flawed experiment has put science on a sidetrack for over one hundred years.

Astronomers know that, with reference to space, where a star is today will not be the same place tomorrow. Stars are in motion through space as is our solar system. This says that if we send a single signal (photon with frequency = 1) directly towards the position where star (x) resides today and it takes the photon 12 years to traverse this distance (12 light years), by the time the photon arrived, star x would have moved.

Now, if we could calculate where star (x) would be in the future, then we could send a signal towards this point in space and if we time it right, it would meet with star (x). But the distance traveled might not be the same. If star (x) is moving away from us then the distance would be greater and if star (x) is moving towards us then the distance would be less. And this brings us to a well known equation (d=rt).

Our signal, being a photon or light if you will, has a constant rate of speed (186,000 mps). So, if r (rate) is constant and d (distance) changes then so must t (time). Its time to bring the space ships into play.

Let us assume that space ship a and spaceship b are exactly 1 light second (186,000 miles) from each other and are moving at exactly the same speed (93,000 mps) and direction (parallel). From what we have discussed above we know that if ship a sends a single directional signal (frequency = 1) towards the position in space where ship b is now, the signal will miss ship b, passing behind it. In order for ship a to reach ship b with its signal, it must create an equation that will calculate the point and time on ship b's path where the signal, coming from ship a, will meet with it. It only has one chance to achieve the encounter with ship b. In addition, ship b will immediately reflect the signal back to ship a. See Figure 10.

FIGURE 10.

Parallel Communications in Space

1. Ship(a) and ship(b) are traveling through space on a parallel course 186,000 miles apart.
2. Ship(a) and ship(b) are both traveling at 1/2 the speed of light (93,000 mps).
3. Ship(a) decides to send a signal to ship(b). The signal will be a single quantum
4. The signal will of course travel at the speed of light (186,000 mps) since it is a light signal.
5. The angle of transmission will be 30 degrees forward of a straight line between the ships at time of transmission.
6. The time to intercept (tti) will be 1.1547 secs.
7. Let us now assume that ship(b) immediately reflects (mirror) our signal back by reversing the process I just showed. This would give us a total round trip time of 2.3094 seconds for the signal.

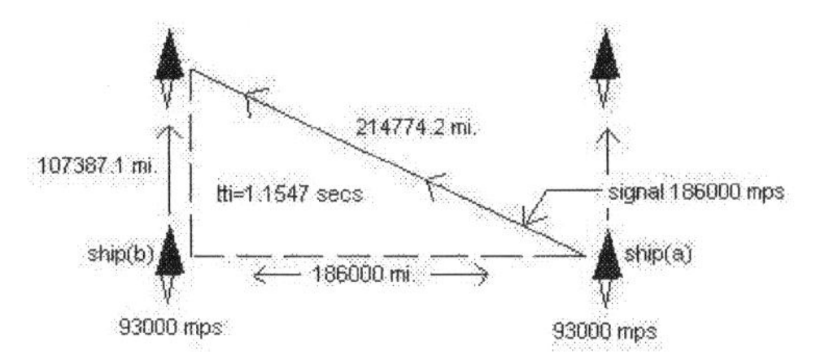

The round trip total time for the signal is 2.3094 seconds and the total distance covered by the signal is 429,548 miles. In the above diagram if you project the signal at 29 degrees it will arrive too late and if you project at 31 degrees it will arrive too early. I would ask those who are reading this work to be sure you understand all the above before proceeding. Let us return to the spaceships.

In this second spatial setup we now have ship a moving through space in a straight line and ship b is out ahead of ship a exactly 186,000 miles but on the same line. Both ships are traveling at ½ the speed of light (93,000 mps). Once again ship a decides to send a signal to ship b, the same signal as before. But the geometry has changed. Before, we had spaceships a and b running parallel and even. Now we have space ships a and b occupying the same line but not the same position on that line.

At the moment when ship a transmits the signal, ship b is 186,000 miles ahead of ship a. Let us look at this a second at a time. After one second the signal has traveled 186,000 miles and arrives at the point in space that ship b occupied, when the signal was originally transmitted. But ship b has moved since then, 93,000 miles to be exact.

In the next second, ship b will have moved another 93,000 miles and this puts its position at two light seconds (372,000 miles) from the point in space where the original transmission occurred. Meanwhile the signal has advanced another 186,000 miles which also puts it 372,000 miles from the point in space where the original transmission occurred. The signal catches ship b at this point.

We now have the signal and ship b in the same place at the same time and the time this took was two seconds. In other words, it took 2 seconds, for the signal transmitted from ship a, to reach ship b. But now comes the return trip for the signal.

Let us assume that ship b immediately reflects (mirrors) the signal back towards ship a. At this 2 second mark, ship b is 372,000 miles from the point in space where the original transmission occurred. At this same 2 second mark, ship a is 186,000 miles from this same point and both ships are still on the same line. This makes the distance between ship a and the signal, at this 2 second mark, 186,000 miles.

Ship a is still moving in the same direction as ship b, while the reflected signal from ship b is headed back toward ship a. At the time of reflection (2 secs.), the distance between the reflected signal and ship a is 186,000 miles. This sets up a configuration where the reflected signal is approaching ship a at 186,000 mps and ship a is approaching the reflected signal at 93,000 mps.

In the first ½ second the reflected signal travels 93,000 or half the distance while ship a travels 46,500 miles or ¼ of the distance. We can see from this that the ratio between the reflected signal and ship a is 2:1 and this speaks of thirds. So we divide up the remaining 46,500 miles giving the reflected signal 31,000 miles while giving ship a 15,500 miles. And finally the reflected signal and ship a meet. Factoring all this in, we find that the point on the line where this meeting occurs is 248,000 miles from where ship a transmitted the original signal. The total distance the signal traveled is 496,000 miles. But what about the time?

We already have the 2 seconds that it took, for the original signal from ship a, to reach ship b. Now we employ the ratio (2:1) we found above. The difference in the rate of approach between the reflected signal and Ship a was 2:1. The distance between them was 186,000 miles. Using these pieces of information we calculate the time to contact to be .666666 seconds.

Now we add the .666666 seconds to the original 2 seconds and this gives us 2.666666 seconds as the original signal's time for the round trip. And the total distance covered by the signal is 496,000 miles. See Figure 11.

FIGURE 11.

INLINE SPACE COMMUNICATION

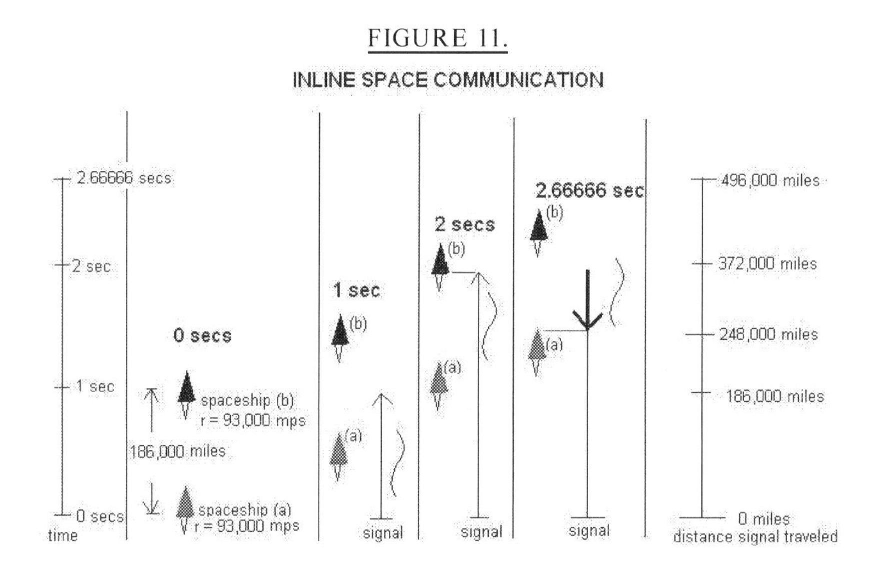

In analyzing Figures 10 and 11 we find that even though the distances between the ships were equal the signal times and signal distances were not. In parallel the signal had to travel 429,548 miles and it took 2.3094 seconds. Inline the signal had to travel 496,000 miles and it took 2.66666 seconds. The magnitudes of difference between Figure 10 and 11 is 66,452 miles and .35726 seconds.

In this third setup we utilize not two but three ships. First we put two of the ships a and b on a parallel course the same as in Figure 11. Next we add another ship c to the first two ships a and b and put it inline the same as before. This gives us two ships parallel and two ships inline within the same setup.

What we are doing here is combining the setups we used for Figures 11 and 12. This leaves only one more change to implement into the setup. I will now replace the ships a, b, and c, with a light splitter/receiver in place of ship a, and reflecting mirrors in place of ships b and c. This will change nothing in the process but the observers.

Splitter/receiver a is parallel to mirror b and inline with mirror c. They are all 186,000 miles apart and are all moving at 93,000 mps, the same as before. We will call this spatial area or platform containing the splitter/receiver a and the mirrors b and c frame (F). Because, they are all in motion at the same velocity through space, all clocks within this setup will measure time equally, when compared to each other.

We will begin by injecting the splitter/receiver a with photons of frequency = 1000. The splitter/receiver a then splits this signal and sends some of the photons (signal x) towards mirror a and some (signal y) towards mirror b. See Figure 12.

FIGURE 12.

Splitter/reciever with mirrors or: The Michelson setup

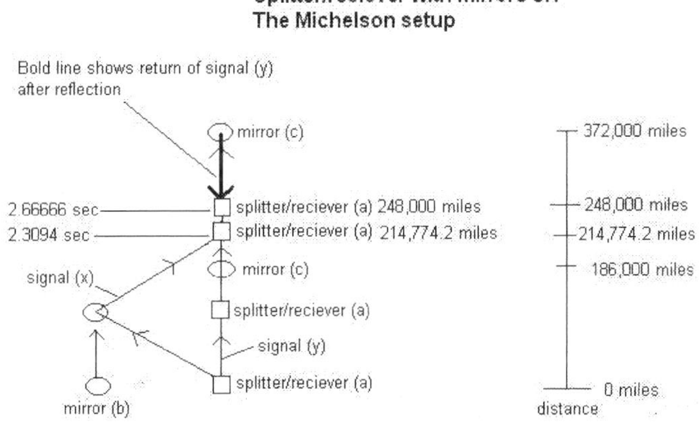

What I have done here is switch my setup with the setup Michelson used in his attempt to detect the aether (space). The only differences between his setup and mine is the distance between the splitter/receiver and the mirrors (186,000 miles vs a few feet), and I gave my setup platform frame F a rate of linear motion (c/2) through space. All he looked for was a difference in speed (without the use of a clock) between the two parts of the beam he split. I looked for times, and distances traveled, during this time

As I said before the equation d=rt is very special. I set my work up on the law known as "the constancy of the speed of light". Within my equations, relating to the speed of the signals, r was always a constant (186,000 mps). Also within my equations, relating to spaceships, splitter/receiver, and mirrors, r was also a constant (93,000 mps) and the distances between the emitter/receivers and the mirrors was a constant (186,000 miles). But what about the frequency shifts of the split signal?

Signal y would have experienced a red shift when it caught up to mirror c at 2 sec. but when it returned to splitter/receiver a at 2.66666 secs., it would experience a comparatively larger blue shift which means it would return to splitter/receiver a, with a blueshift. Signal x would have experienced a red shift when it arrived at mirror b and an additional red shift when it arrived back at the splitter/receiver a. This says that the return frequencies of signal x and signal y would have an appreciable difference between them when recombined at splitter/ receiver a.

In conclusion: The above work was put together by using mathematics and the laws of motion, along with the conclusion that space transmits light through it at a constant velocity. I see two major flaws in Michelson's experiment. First, his mirrors were just a few feet apart, and the thing he was measuring moved at 186,000 miles per second. And secondly, Michelson left time out of his results. As you can see from the work above, time and distance can change the story. Michelson left time out of an experiment involving motion and he also inserted a physical factor (light) which is undefined as to its properties. This experiment was flawed from the start.

Let us now look at Fact 1 again and how it connects to what we call the Doppler shift. Fact 1 says, "If it takes the light ray more time to catch you when you are in a motion ahead of the advancing ray, then it also takes the light ray more time to pass you." The first thing to be done here is create an experimental setup using frame (S) as our inertial frame of reference. Remember, frame S is the coordinate frame to the imaginary frame **(^)** we discussed before.

We will name the following setup (event 1). We have an observer O1 with a clock C1 setting motionless at frame (S) and a light source L, 186,000 miles away from frame (S) and also setting motionless with reference to frame (S). Source L sends a single particle/wave of light towards observer O1. We will arbitrarily make the length (period) of the light wave 186,000 miles in order to simplify the math. (See Figure 13.)

FIGURE 13.

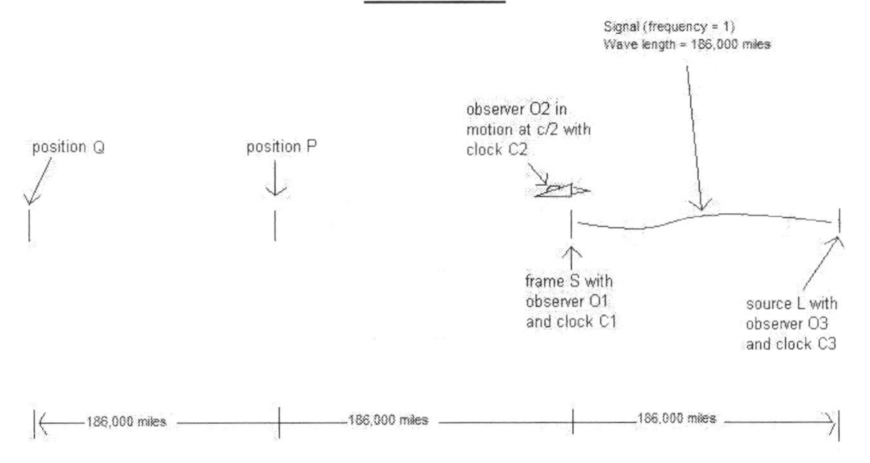

The fact that I chose such a long wave length will not in any way affect the outcome of this particular work since it is understood that all of the particle/wave travels at the same speed c. I could also say that we attach a 186,000 mile long string to the particle/wave and it would make no difference, in relation to the results we're after.

The rate of approach, with reference to the light wave approaching observer O1, is 186,000 mps. Here we know that the speed of light is constant and assume, relative to clocks on earth, that the true speed of light is 186,000 mps.

With reference to observer O1 and his clock C1 at frame (S), it will take the head of the wave exactly 1 sec. to arrive at observer O1's position at frame (S). It will then take an additional second, with reference to observer O1's clock C1, for the tail of the light wave to arrive at observer O1's position at frame (S). This means, it took a total of 2 seconds, with reference to observer O1 and his clock C1, for the light wave to travel from source L and pass him at frame (S). This also means, it took 1 second, with reference to observer O1 and his clock C1, for the complete light wave to pass him at frame (S)

Take note here that no Doppler shift (lengthening of the wave) resulted from this event 1. The frequency of the light wave when it was emitted from source L was 1 and the frequency of the light wave when it passed observer O1 at frame (S) was still 1. Let us now set things in motion.

In this setup (event 2), we put an observer O2 and his clock C2 with observer O1 at frame (S) and an observer O3 with his clock C3 at the source L. When the same 186,000 mile long light wave, as in event 1, is emitted from source L, we simultaneously put observer O2 in motion at c/2, away from the light wave which is approaching from source L.

According to relativity, observer O2's clock C2, once observer O2 is put into motion, will run slow to some degree, when compared to observer O1's clock C1 at frame (S) and also when compared to observer O3's clock C3 at source L.

We are now at the point within this event 2, where we apply Fact 1 which says, "If it takes the light ray more time to catch you when you are in a motion ahead of the advancing ray, then it also takes the light ray more time to pass you. We only need to substitute the words light wave for the word light ray to make this change complete. Fact 1 now says, "If it takes the light wave more time to catch you when you are in a motion ahead of the advancing light wave, then it also takes the light wave more time to pass you."

Remember, it was Dr. Einstein himself (page 25, section XI in his book RELATIVITY-The Special and the General Theory, who set up the foundation for Fact 1, when he said, ""Just when the flashes of lightning occur, as judged from the embankment, this point M1 naturally coincides with the point M, but it moves towards the right in the diagram with the velocity v of the train. If an observer sitting in the position M1 in the train did not possess this velocity, then he would remain permanently at M, and the light rays emitted by the flashes of lightning A and B would reach him simultaneously, i.e. they would meet just where he is situated. Now in reality (considered with reference to the railway embankment) he is hastening towards the beam of light coming from B, whilst he is riding on ahead of the beam of light coming from A. Hence the observer will see the beam of light emitted from B earlier than he will see that emitted from A.""

Let me repeat Fact 1, "If it takes the light wave more time to catch you when you are in motion ahead of the advancing light wave then, (according to Dr. Einstein above), it will also take the light wave more time to pass you." Because of its importance, I would ask the reader to be sure he/she understands what was just said, before continuing.

Now let us bring into this setup for event 2, another of Dr. Einstein's laws that is connected with the constancy of the speed of light through space. On page 229 in the book, Ideas and Opinions, by Albert Einstein, he says, ""The second principle, on which the special theory of relativity rests, is the, "principle of the constant velocity of light in vaccuo." This principle asserts that light in vaccuo always has a definite velocity of propagation (independent of the state of motion of the observer or of the source of the light)."" From this we must conclude, If the speed of light is independent of the motion of an observer then it must also be independent of the motion of the observer's clock. This conclusion presents a mathematical problem relating to the equation d=rt.

It will become obvious once we carry out event 2, that Fact 1 above and the part of the Law for the Constancy of the Speed of Light (the motion of the observer), are in direct conflict. Light at speed c, leaving source L and traveling for one second will travel some distance. But we also know that the measure of this one second is variable. Dr. Einstein says this himself. This says mathematically; the distance an object travels in one second, with reference to a clock in motion at velocity v, will necessarily be a shorter distance than the distance this same object will travel in one second, with reference to a clock in motion at velocity v^2. Let us contemplate all this for a moment.

It is important here that we keep in mind, the speed of a particle/wave of light is in fact, a constant, with reference to space, just as Dr. Einstein said. And, as we discussed earlier in this work, the source is nothing more than some object of mass (galaxy, star, filament in a light bulb, etc.), which is simply made of particle/waves of light running in circles.

Hence, we understand why the speed of light is independent of the "spatial" motion of the mass source as a whole. This says, the speed of a particle/wave of light is independent of the spatial motion of its mass source as a whole, but not to the prior motion of itself, before it was emitted from the source.

A car in motion after darkness begins (night,) turns on its headlamps. What happens is as follows: The outer electrons within the atoms making up the filaments within the headlamps, are put into motion (accelerated), due to a voltage being applied to the filaments. This causes the outer electrons, through their added energy of motion (acceleration), to escape the atom they were within and jump to another atom. This accelerative force causes a contraction of the electron, which effects the spatial symmetry of the active forces within the electron (quanta traveling in orbits). Hence, particle/waves of light (photons) are forced out of the electron. Therefore, the particle/waves of light within the photon emitted, have always been in motion at speed c.

Let me restate, "The speed of a particle/wave of light, is independent of the "spatial" motion of its mass source, as a whole, but not to the prior motion of itself, before it was emitted from its mass source". This part of the discussion only addressed the principle relating to the source's relationship to the constancy of the speed of light through space. The other principle, which relates to the observer, needs a little more explanation. Let us now return to event 2.

As I said before, event 2 will have the same setup as we used in event 1 (Figure 13.) with the exceptions, we added two observers O2 and O3. Observer O3 is placed at the source L which is a co-ordinate of frame (S), hence observers O1 and O3's respective clocks will keep equal time throughout event 2. Observer O2's clock C2 will be in motion throughout event 2, hence, clock C2 will be keeping time, in reference to event 2, at a slower rate ($(1/\sqrt{(1-v^2/c^2)})$) than will clocks C1 and C3. Let us now set all clocks C1, C2 and C3 to (t = 0) and begin event 2.

The 186,000 mile long light wave is emitted from source L, and at this same moment ((simultaneously with reference to frame (S)), observer O2 is put into motion at speed c/2 in a direction ahead of the light wave.

As before in event 1, the head of the light wave reaches observer O1 at frame (S) in 1 sec, with reference to clock C1 at frame (S). Meanwhile observer 2, who is in motion ahead of the light wave at c/2, is 93,000 miles in advance of the head of the light wave which has just reached observer O1 at frame (S).

When we advance our event time another second with reference to observer O1's clock C1 at frame (S), we find that the light wave has traveled another 186,000 miles and arrives at position P, while observer O2 has traveled another 93,000 miles and also arrives at position P. Hence, the head of the light wave overtakes observer O2 in 2 secs., in reference to clock C1 at frame (S). Meanwhile, the tail of the light wave arrives at observer O1's location at frame (S) at this same 2 sec. period in time, relative to observer O1's clock C1 at frame (S).

At this same point, let us compare data between observer O1 and observer O2. As we found above, observer O1's clock C1 shows 2 seconds for the time it takes for the head of the light wave to catch observer O2. But relative to observer O2, we find a difference in times. Observer O2, who has been in motion at c/2, is experiencing a slowing in his clock's time keeping. Hence, observer O2 will record less than 2 seconds for this same action. Follow here.

Observer O2 sees that he has arrived at position P, which was 186,000 miles from where he started, and he knows he was in motion at c/2, with reference to his starting point S and he also knows that the light wave originated 186,000 miles beyond point S at source L. Observer O2 then plugs this data into the equation d = rt and finds that the calculated time does not agree with the time recorded by his clock C2. Observer O2's clock C2 shows a time (1.732 seconds) which is less than 2 seconds, for this particular part of event 2.

Once again, let us advance the event time by 1 sec., with reference to clock C1 at frame (S). The head of the light wave has advanced another 186,000 miles beyond position P and arrives at position Q, while observer O2 has advanced 93,000 miles beyond position P (half way to position Q). As you can see, only half of the light wave has passed observer O2 during this same 1 sec, with reference to clock C1 at frame (S).

This leaves another 93,000 miles of light wave to pass observer O2, who is still in motion at c/2, which will add another second of time with reference to clock C1 at frame (S). Finally, just as observer O2 reaches position Q, the tail of the light wave passes him and he stops his clock C2, while observers O1 and O3 stop their clocks also.

With reference to clock C1 positioned at frame (S), it took two seconds for the complete light wave of 186,000 miles to pass observer O2, once the wave caught up to him, while observer O2 was in motion at speed c/2. But once again we find, observer O2's clock C2 shows less time (1.732 seconds) for this part of event 2 than does observer O1's clock C1.

In the end, observer O1's clock C1 shows 4 seconds elapsed during all of event 2, while observer O2's clock C2 shows less than 4 seconds (3.464 seconds) elapsed. Observer O3 at source L with clock C3, because he was at rest relative to observer O1 at frame S, also records 4 seconds for all of event 2, the same as observer O1 with clock C1 at frame S.

The end result here is, "In reference to a clock C3, placed at the source L of the light wave, it will take a greater amount of time for the light wave to overtake and pass, observer O2, in motion away from frame S, than it would if observer O2 were stationary at frame S. This agrees with Fact 1.

But notice that observer O2's findings, with reference to his clock C2, shows that once the head of the light wave caught him at position P, and the tail of the wave passed him at position Q, it took less than 2 seconds for the complete 186,000 mile wave to pass him while he was in motion at c/2. These facts force observer O2 to calculate (d=rt), that the speed of the particle/wave, as it passed him, with reference to his clock_C2, was greater (209,932 mps) than 186,000 mps c. Hence, observer O2 will be mathematically forced, to record a frequency greater than one for the light that passes him.

Since we know (empirical evidence) that space transmits light at a constant velocity through it, then the results found here by observer O2 and his clock C2, are in direct violation of the principle that says, the speed of light is independent of the motion (clock) of any observer.

This is why the reference frame **(^)**, (introduced earlier) is so important. It separates the truth from the possibilities. This means observer O1 on earth, earlier within this work, was correct when he stated that the spaceship passed him at c/2 and observer O2 on the space ship was wrong when he stated that the earth was passing him at c/2.

There is only one "mathematical relationship of motions", that will explain the results connected to event 2. The rate of approach between the light wave and observer 2 had to be 93,000 mps and not 186,000 mps. In other words, the speed of the light approaching observer O2, with reference to observer O2's motion, had to be 93,000 mps and not 186,000 mps, which is dictated by the principle (motion of the observer) we are addressing.

This relationship of motions (approach rate equal to 93,000 mps) agrees fully with Fact 1, and also with observations (clock times) made by the independent observers O1 and O3, but conflicts with the part of the postulate for the speed of light which says, the speed of light is independent of the motion of any observer.

Remember before, near the beginning of this work where we discussed how the speed of light Dr. Einstein referenced, would only be in accordance with seconds measured by clocks on earth, which we determined would be running slow due to earth's motion through space. We now see, from what I have just shown, only an observer positioned at frame S, which is a coordinate frame to imaginary frame **(^)**, can measure the true speed of light in vaccuo (through space). I will now show another work which I hope, will shed a little more light on our problem above.

THE LORENTZ/FITZGERALD CONTRACTION

The length of time it takes for the particles of light and their associated kinetic waves to travel in their orbits within the quarks determines the frequency of the resultant matter waves around the quarks. The frequency, with which the matter waves propagate, determines the rate of molecular action for the object of mass they reside within and this frequency is directly tied to the orbit period of the particles of light making up the quarks.

When an object is "hypothetically" stationary in space, the particles of light, within the quarks, travel in their orbits at speed c. This causes the "angular" matter waves created by these quarks to propagate outward at a frequency that can be calculated by combining, the time it takes for the light particles to complete their orbits, with the speed at which they are moving thru space itself.

As I have said elsewhere within this work, this matter wave frequency, I have associated with the quarks, can vary, depending upon the motion through space, of the objects of mass that the quarks are within. This is the mechanical basis for the phenomenon we call the Lorentz Contraction.

As an object moves through space the particles of light within the quarks, making up this object, find that they cannot travel in their orbits with the same speed as before. This is because they are forced to share their speed, within their orbits, with the speed of the object they reside within, which is also moving through space. This slowing, in the time it takes, for the particles of light to complete their orbits in the quarks, causes the matter waves associated with the quarks, to propagate at a slower frequency and with diminished strength. Since we know that the rate of molecular action is determined by the speed, with which atoms interact with each other, we now know why this interaction rate is reduced as an object is put into motion. Hence, the clock for this object runs slower. Be mindful that light speed was never less than c above.

In addition, as a body of mass is accelerated the particles of light within the quarks making up the body of mass move about their orbits in a longer period of time. Hence; their matter wave frequency is slower and their matter wave strength is also weaker, this causes the particles making up the atoms making up the body of mass to move closer together due to their matter waves being diminished in their role as separators (quantum force). This action also causes the electrons to take up orbits closer to the nucleus for the same reason. An object under motion through space, along with its diminished matter waves, is what generates the transformation on mass matter we call the Lorentz contraction. Basically; the faster an object of mass moves thru space the shorter and slimmer it becomes. And if we were to ever cause the object of mass to attain the speed of light we would necessarily create a photon.

Let me reiterate that universal time ("NOW") is not effected at all, but only the timepiece (the clock). Only by slowing the speed with which light moves c, could we affect universal time (NOW). Discounting the use of mediums, the two situations in which the speed of light would vary from c would be right after the Big Bang and when space reabsorbs the individual particles of light after they have been separated from their photons. With that being stated, I will also hazard a guess that if we could freeze a vacuumed cylinder to as close to absolute zero as possible, light just might move slower than normal through the cold space within this same cylinder. The speed of a particle of light is directly related to the pressure being applied to it by the space surrounding it.

PART THREE – TIME

TIME

What exactly is time? Time is one of the most misunderstood ideas man has ever conceived. Some say that time flows. Some say it simply goes by. All descriptions are based on one premise which is motion. To put it simply, time is a man made measuring tool for gauging the motional happenings within the Universe around him. Let me tell you a story if you will.

Once upon a time there were two cave men (Morg and Borg) from two different clans who met in order to go on a hunt. As they preceded on the hunt Morg ask how Borg's wife was. Borg replied by saying his wife had died. Morg then ask Borg when was it that his wife died. Borg replied that it was about three moons and three straight ups after the bear eat Torg. Morg, coming from a distant land, did not know Torg was eaten by a bear. So now Morg wanted to know when Torg was eaten by a bear. Borg replied that that happened two moons and five straight ups after the top blew off the mountain. Now everyone remembered when the volcano erupted. This allowed Morg to finally build an understanding as to when all these events occurred. The reason for the confusement was because their clans had no time system with a solid reference point. You have to understand here that they had yet to develop a time system with such components as Eons, Centuries, Years, Months, Days, Hours, Minutes and Seconds etc. Big events such as earthquakes or volcanoes erupting were possibly how they originally gauged things that happened within their local tribes. Note that the western world still follows this pattern, BC (Before Christ) and AD (After Christ). Now back to the story.

Finally the hunt was over and it came the time when Morg and Borg would part their ways and go back to their respective clans. They decided that they wanted to do this again in the future but now they had to decide just when. Morg held up five fingers and then pointed straight up which meant that they would meet five days from now at noon when the sun was straight up.

This time system that the cavemen used worked for their purposes at that time but would be found lacking in the present world. The way they developed their system is obvious. In order to develop a reliable time system you must first have a natural occurring event which is both periodic and repeating to base your system on. Back in the caveman time the event which was most noticeable as periodic and repeating was the sun coming up and going down. Next came the dark and full of the moon. In the deep northern and deep southern hemispheres the seasons were probably a part of their time systems. At any rate I think you can understand what I'm trying to convey. If the Universe had no motion within it then there would be no time because our time system, at its foundation, is based upon movements within the Universe itself.

I know you have heard of the supposed paradox where one twin boarded a space ship which traveled at near the speed of light while the other twin stayed on earth. After 50 years the first twin returned to find his brother an old man while he himself hadn't aged a day. But we must realize that the Law of the Conservation of Energy holds even for time itself because time, at its base is directly tied to motion. The first twin did not somehow slow down universal time but only his bodies internal interactions. This says; under motion you will age at a slower rate because the rate of interaction going on between your body's quarks, atoms and molecules will become slower the faster you move thru the fabric of space surrounding you.

While I am on the subject of manmade systems let us talk a little on the subject of dimensions. Dimensions are a manmade system used to describe things in three axis using the terms height, width and depth. The Universe doesn't separate itself into different configurations of being. It simply is. When you hear of 4 or even 27 dimensions, remember just what dimensions really are so as not to be mislead. Using more than 3 dimensions to describe something is just another attempt to mathematically create a mathematical magic wand. Ten dimensional string theory arose from just such a scenario. If you look at this logically, just as a line can have infinite points a sphere can have infinite dimensions. To those who read this work in the 31 century; make sure all non-fictional ideas such as this are scrubbed from science books. It just causes confusion for those trying to learn the skill of physics. Now back to our original subject which is time and motion.

Some in the scientific community have tried to treat time as the fourth dimension. Time, as I said before, can be defined as process used to measure movement within the Universe. If all the parts of the Universe stood still with no movement, there would be no time. But, since all the parts of the Universe do move in one way or another then an apt description of this activity would be; "SOMETHING MOVING". When an object moves it is said to gain momentum. Since the Universe as a whole contains movement everywhere then the term Universal momentum and time must necessarily be synonymous. I hope you enjoyed the little story of cave man time I spun for you. Let us continue.

To repeat; time is simply a man made system, which he uses to define and measure events happening in the world around him. Some of the time scales he has constructed include years, months, days, hours, minutes, seconds, centuries, eons, etc. The one common denominator within all scales or systems of time is movement.

Many within the scientific community believe that there was no time (movement) before the Big Bang. I, along with the Second Law of Thermodynamics, say that this belief is erroneous. In fact, within an infinite universe, there would be many Big Bangs, which makes our Big Bang just a local event. Logically, Hoyle's steady state theory must eventually rise from the ashes! Note: a perpetual "non-chaotic" universe must, at its foundation, possess a balanced repeating symmetry of its internal forces and how they interact. I am right on this!

We must remember that the Second Law of Thermodynamics implies, "If the car (universe) runs out of gas (available energy) and stops, it will remain stopped unless more available energy is added". The fact, that I am typing this document into my computer at present, convinces me that the universe has, and always will, contain movement (time).

This model dictates that all matter to include even space, is made of particles of light and light, as we have found, has only one speed, at least within our local area of space, 186,000 miles/sec. The constancy of the speed of light is ultimately dependent upon the characteristics of the space around it, and I need to qualify my previous statement to reflect this. As I present this model and speak of light particles coming to rest, then transforming into the particles making up space I need to make it clear, the particle of light has discontinued its motion through space and now expands which adds more pressure to the fabric making up space itself. This is the condition where a particle of light becomes dormant (inert) energy and becomes part of the fabric of space.

Now we have established the rate of motion the universe uses as it changes from the past into the future, which is the speed of light. This we will call Universal time. Let us now turn to Local time.

The Milky Way, along with the rest of the matter ejected from the Big Bang, is moving at some speed away from the original location in space where the Big Bang occurred. Current theory holds that both space and time were created simultaneously from the Big Bang. I found that I could not include this view within my model for it has "Something" being created from "Nothing", which is a direct violation of all known and accepted Laws of Physics. In addition, Axiom # 2 also disallows the universe to have ever existed in such a configuration. My model holds that space simply ruptured due to pressure. This view is much simpler and most importantly, it is possible. No other theory offers what mine does which is a plausible perpetual return manifold for rejuvenating the universe itself.

The speed, at which we move through space, causes a slowing of the molecular action within us (Twins Paradox). This means that our time measuring devices will run slower than for someone positioned close to where the Big Bang occurred. (See the Lorentz Contraction)

An interesting point to be made here is; we are moving away from the area in space where the Big Bang occurred, at a certain speed. If we sent a space ship back towards this Big Bang area in space at exactly the same speed the earth is moving away from it, the clocks on this space ship would speed up even though, relative to the earth and accepted physical laws, the space ship's clocks should be going slower. The reason for this becomes understandable once we remember that the earth is moving at some velocity, away from the area of space where the Big Bang occurred, and as such, its clocks are running slow.

In fact, our space ship will never reach this area we intended to send it to. Relative to the earth the space ship is going in the opposite direction that the earth is moving and at an equal speed to the earth. Relative to the point in space where the Big Bang occurred, this space ship is standing still. What we have actually done with our space ship is reverse the Lorentz Contraction in this instance. This space ship will expand and its clock will run faster, and if humans are aboard, they will age faster. This hypothesis, when subjected to empirical endeavors at some point in the future will prove me correct if man can survive that long.

Again; the Milky Way is moving at some speed away from the original location in space where the Big Bang occurred. The speed, at which we move through space, away from this central location, causes a slowing of the molecular motion within us (Twins Paradox). This means that time (subatomic and molecular action) for us is necessarily slower than for someone positioned close to where the Big Bang occurred. One of the reasons I will give for postulating that all clocks on earth are running slow is that, a space ship taking off from earth and attaining a speed equal to 80% of the speed of light, experiences a slowing of its clocks by 40% (approx.). I only assume the accuracy of this calculation, from various data I have read. But in actuality, the correctness of the data used in this calculation, neither validates nor invalidates the idea being presented. This is because the earth is not standing still in space, but is continually orbiting the sun, which moves about the galaxy, which together with the other galaxies (local group), are moving away from the position in space where the Big Bang occurred, as is the rest of the matter that was ejected from this event. If the earth was in a stationary position at the point in space where the Big Bang occurred, then the clocks on the space ship, which is traveling at 80% the speed of light would, by the clocks on earth, be running approx. 80% slow.

At this point I need to point out that, if the Lorentz Transformation is valid and space is actually void or empty, then we have a situation where an object must "literally" know it is in motion in order to effect the changes we see it undergo. If we have a spaceship in motion at 99% the speed of light far out in space and sufficiently removed from any massive objects, we calculate, with reference to earth, that this spaceship's time will be dilated by some degree. But this leads us to question how motion, through empty space, can cause time dilation.

Let us now use reverse logic in this instance to further analyze this situation. In reference to earth, this spaceship's clocks or time, has slowed because it is in motion away from earth at 99% c. But the people on the spaceship don't recognize this and think everything is progressing as normal. In fact, if we let these people forget about earth altogether, they will think they are stationary and the earth is in motion away from them at 99% c. And according to the Theory of Relativity, they are equally right with reference to the spaceship and the ignorance of the people on the space ship.

So now the Captain, not realizing he is already in motion at 99% c, in reference to the earth, decides to send a probe out, which accelerates up to a velocity equal to what he measures the earth to be receding from him. The Captain now has the earth receding from him in one direction at 99% c and his probe receding from him in the opposite direction at 99% c. This leads to a situation where mass, with reference to the earth, is in motion at 198% c. According to ideas formulated by the Relativity Theory, there is no problem with this situation with reference to the space ship. But if we return to the reference frame associated with the earth, we now see the Theory of Relativity utterly run over itself. The point to be shown here is, if motion through space can cause time dilation, it can only be space, which is ultimately responsible for this phenomenon, and empty space cannot, under any circumstances, cause an effect such as this. With this being said, I need to make it clear that I do not doubt the Lorentz Transformation but am questioning, the idea of an empty space, along with some of the ideas that have evolved from the Theory of Relativity.

Some will wonder where I am going so I will get to the main point. All motion and time within our local area of space can be collectively compared by use of a mutual frame of reference (area in space where the Big Bang occurred). This is the concept that will allow us to actually have a common and "True Time" reference point with which to mutually calibrate all clocks. This would allow local time for any event happening around us, to be adjusted to what I will call "space time" so that no matter the reference frame of other observers they can all obtain mutual measurements of the events in question.

I realize that this does not agree with Relativity but we must also remember that Einstein formulated his concepts without a physical space in the equations. In the example given by Einstein where an object is dropped from a moving railway carriage, the statement is made, that different measurements by different observers in different motion are equally correct for this event. This is true only so long as their measurements are compared to their inertial frames of reference and they make no judgment as to the actual distance through space the object traveled. Only by calculating the motion and direction with which they are moving through space, into the equation, can they compute the correct distance and time that it took for the object to travel through space during this event. Space is the "one true" inertial frame of reference for any and all measurements of motion or time. For you students in the year 3000 AD; up until now (2016 AD), I've never made a penny for my work but I can truthfully say," I've loved every minute of it". Believe it or not as you wish!

Local time works well for most time keeping efforts here on earth but in the future when we are trying to relate events to other worlds or to space vehicles moving at various degrees of the speed of light, "space time" will be a necessity.

If we place ourselves out in space a sufficient distance so that we could (hypothetically) see all the kinetic energy that blew out of the Big Bang, what we would see is a huge doughnut spread out on one plane similar to our galaxy's configuration, only with a huge hole in the middle. We would also notice that this doughnut is expanding. We must also note that all this kinetic energy is expanding away from a central point, and this fact is central to my views. Some schools of thought hold, that the gravitational field connected with all this matter ejected from the Big Bang, will eventually draw all this matter back in and this will generate another Big Bang. This model shows this concept to be unattainable as the doughnut configuration allows space to replenish the center of the doughnut with little or no resistance from the matter expansion. Also, even if we give the expansion a spherical configuration, the gravitational characteristics for the inner area of this sphere will be the same as the doughnut. This says that the output from the Big Bang will remain in a state of expansion. Remember, "Light travels faster than mass." Note: space can and does replenish itself else electromagnetic fields could not exist. But that's another story.

Let us now give two men, whose names are a and b, identical clocks, which we have calibrated to the same time. We place a and his clock at the central point of this doughnut (area in space where the Big Bang occurred) and place b out on earth with his clock. What will immediately happen is that b's clock will slow down compared to a's clock, though b will be unable to see this as his internal biological clock (molecular action) also slows down. This is due to the slowing of all the molecular action on earth and this slowing is produced by the atom's reaction to moving through space. See "The Lorentz Contraction".

Space itself, within this model, has undulations of pressure moving through it. We must understand that, within an infinite-perpetual universe, there is no absolute non-moving frame of reference. But this central point in this doughnut, from where all matter was ejected from space (Big Bang), is the next best thing. We do not need an absolute non-moving frame of reference as long as we realize that this doughnut and all its kinetic energy, have a reference point which allows all kinetic energy, and the objects it makes up, to have a mutual central time (space time). Basically, we can say that since everything in this doughnut is moving away from this central point, then the basic reference to local time for the whole doughnut would have to be this central point.

The ultimate point to be made is, local time (molecular motion) for you can be different than for me, if our rate of movement through space is different. But no matter the difference in molecular actions or difference in speed between objects anywhere within the universe, no object or body has the ability to exist outside of "Now".

Some hold the assumption; it might be possible to travel back in time, if you could somehow travel at a speed in excess of c. I need to point out, in reality; the past no longer exists. What the past was, has changed into now while undergoing change in order to generate the future. Put into simple terms, the past and future are both right here with us and it all exists simultaneously in what we call "Now".

I will postulate: Time (now) does not undergo change discontinuously or disjointedly. The basic foundation for time is directly connected to all the motion within the universe itself. Time, it must be understood, is a man made idea (gauge) to be used for measuring actions happening around us in nature.

To those of you in the year 3000 AD, motion does more than affect our measurements of time. Without motion there would be no time. Hawking stated that time and space did not exist before the Big Bang. I hold that waves of pressure moving thru space allows for a form of time before our Big Bang and of course all the other Big Bangs possessed movement hence, time would exist there also. Note that, since we share mutual space with the output from all other Big Bangs time has always been present within the whole. As for space, which is presently seen as a void and not existing before the Big Bang, how does one expand let alone create this empty space.

We, as humans like to gage or measure the things and events happening around us in nature as if they were in a static or non-moving frame of reference. This is just one example of why the classical measurement method, is limited in its ability to describe reality in an infinite-perpetual universe. The fact that we can never truly know all the movements within an infinite and perpetual universe should not stop us from creating what would be the next best frame of reference. Again; this frame of reference would be back at the position in space where the Big Bang occurred.

Basically, if we could calculate our exact rate of motion through space with reference to the position in space where the Big Bang occurred we could calculate just how slow our clocks, and the clocks of the whole doughnut (universe), are running.

Not only that, we could overcome the problem of local time for two people describing the same event from different places, while moving at different rates. The only addition to their calculations would be the difference in their individual motions and distances from the event and then factoring in the value for this mutual time reference. What I am saying here, there is a local time for all objects, which were created from the Big Bang. All that needs to be done is to figure out where in space the Big Bang occurred and the rate at which we are moving away from this place.

From the data I've seen, I will try to give a close approximation of the rate at which the earth is moving away from this Big Bang area in space. If the space ship, which is traveling at 80% the speed of light, has a time dilation of only 40%, referenced to earth, then the earth is moving at a speed, away from the Big Bang area in space, at approx. 40% the speed of light. Hence, our clocks (rate of molecular motion) are running approx. 40% slow. This means, the quark matter waves (quantum force) within us should be propagating at a frequency directly related to an angular velocity equal to 111,600 miles per second.

This calculation assumes the data, stating that an object moving at 80% the speed of light will have a time dilation of 40% when compared to a clock on earth, is correct. This would show our time (molecular action) dilation to an accuracy of plus or minus 5%. This calculation assumes that the quark matter wave propagation frequency, of an object, located in the area of space where the Big Bang occurred, would have a matter wave propagation frequency, directly related to the speed of light. This simply means, the particles of light traveling in orbits to make up the quarks, while located in the area of space where the Big Bang occurred, would have an orbit speed equal to the speed of light. And the propagation frequency would be factored by this speed divided by the distance around the orbit.

On the other hand, if this object was moving away from this same point in space at the speed of light, its quark matter wave propagation frequency would be zero. It would turn into light, which has no angular matter wave associated with it. Since this "space time" probably won't be needed until we start real space travel, let us get on to other things.

If we can determine the rate at which the earth is moving away from this central point, and measure the distance between the kinetic matter waves being propagated from a quark, we can then calculate the size of quarks. This will be the beginning of true subatomic exploration. This quark's mass force characteristics will show up everywhere in nature around us.

As an example, let us assume the quark's diameter to be approx. 10^{-15} centimeters and the earth's velocity through space to equal 119,735 kps or 40% c. This gives us a calculation that has the particles of light traveling 179,600 kps as they travel through their orbits within the quarks. We now figure in the diameter for these orbits, which is 10^{-15} centimeters and we then calculate a circumference of 3.1416^{-15} centimeters. This distance divided by the rate 179,600 kps, gives us a frequency of angular matter wave propagation equal, to one wave every 1.75^{-20} sec. Every type of mass wave (force) in nature around us would have "approximately" this basic frequency inherent to them. All the values I use above are hypothetical but the idea behind them is anything but. If we were to take an electron and analyze its matter wave we would find a very small wave frequency within it, similar to what I've postulated above. This would be a possible proof in my view that particles of light in orbits around each other make up what we call quarks.

The main point, in all that I have presented concerning time, is that time is a man-made gage for movement, and the movement for time within the universe has but one rate, the speed of light. The rate at which "Now" moves, from the past into the future is equal to the speed of light c. And all matter moves at the speed of light, for even when we think we are standing still, the particles of light internally making us up are still maintaining a rate of motion equal to c. I need to point out here that matter waves proliferate at a speed less than c, but remember that matter waves, within this model, are simply a reaction by physical space to particles of light moving through it. **Matter waves are basically space waves and we, as humans, are also basically space waves**.

RELATIVE TIME

Time, as a noun, is defined by the Encyclopedia Britannica as: "A non-spatial continuum that is measured in terms of events which succeed one another from past through present to future." It becomes clear that within this definition we see past, present and future as three distinct factors that combine, within a successive order, to generate what is called, the continuum of time. But let us analyze this definition more closely.

The past is seen as, that which has already occurred. The present is seen as, that which is occurring at present or "now" if you will. The future is seen as, that which has yet to occur. But we must realize that this description, with reference to reality, will not hold up.

The past is occurring even as we speak and so too is the future. This means that both the past and the future are being generated, in reality, from the point in time known as "now". Some might question this, but if we can imagine a beginning in time, then we would have an original "now", when neither the past or future existed, as of yet. This allows me to say that, "now is real in nature around us, and past and future are man-made ideas relating to "nows" that were, and" nows" that are yet to come." With reference to time itself, we cannot treat the past and future as separate and distinct from the present.

All the energy, of what we call the past, is still here and now, being used by the present (now), and the future cannot become reality without the changes made, by the energy of now. This idea is backed by the conservation laws of mass and energy.

Under analysis, this says that the idea of time travel is unattainable as both the past and future are here and now. In other words, "How could an object, hypothetically in motion at a speed greater than c, reverse all the momentum of the universe around it?" In addition, if one were to jump out of the present (now), which contains both the past and future (all the energy that has ever existed), where would he go?

It must be understood; if someone asks you for the time, you will give them a man-made measure connected to the birth of Christ, millenniums, centuries, years, days, hours, minutes and seconds. But if we ask the universe for the time, the answer will always be "Now". This is the future, this is the past, this is now!

But then, what is this "now" we speak of here? "Now", is a constantly happening energy-transformation, for the universe as a whole. Hence, we see that time is directly tied to motion. Follow here; Let us hypothetically stop all motion in the universe. We now ask, how would this affect time?

In analyzing this event we need to bring into play the equation (d = rt) and change its priority to r = d/t. With all finite parts of the universe, in a true state of rest (stopped) as we hypothesized above, let us plug the results into our equation. First, with no movement (r = 0), there can be no distance (d) attained or covered (displacement). Second, with no movement (r = 0), there can be no velocity (r) attained. And third, with no movement (r = 0), all clocks, conventional and atomic, would stop or t = 0.

The equation r = d/t, when broken down, and with reference to the above, simply says; If an object is put into motion (r), it automatically covers some distance (d), in some time (t). In other words, in a stopped universe you could have distance between objects but you would be unable to traverse these distances because motion (time) ceased to exist. From the above, we can plainly see that if motion does not exist then time does not exist. Hence, time is a result of motion. But, we ask, motion of what?

With reference to the senses of the human mind, the point in time known as now, is an ever-present side effect generated by the motion c of all the energy and matter in the universe around us. Seen in this light, "now" is simply an effect we feel as the universe constantly changes (transforms) around us. Time is a man made system, used to measure the relative aspects of the motions happening around us during what we call "now". "Now" is an effect of energy in motion. Hence "now" would have a transformation rate equal to the speed of light. And all this taken together says, the universe is transforming at the speed of light. Everything contains kinetic energy now!

Our minds and bodies as well as all matter, feel and react to this effect (now), and depending upon the rate of motion we are in, we will feel and react to this effect differently. What I speak of here, is that our measures of the universal time (now) can be different, depending upon our rates of motion through the space around us. This result becomes a variable factor I call local time. Dr. Einstein called it relative time.

On page 37 in the book, RELATIVITY- The Special and the General Theory, by Albert Einstein, he says, "As a consequence of its motion the clock goes more slowly than when at rest." He derives this result from applications involving the equations relating to the Lorentz Transformation.

Using the Twins Paradox as an example; if your twin left on a spaceship, whose velocity was 99% c while you stayed on earth, and your twin returned 40 years later, you would have aged 40 years while your twin would seem to have aged not at all.

With reference to earth, you and your twin are both 40 years older but his molecular action has been slowed during this 40-year period. If, before your twin left on this 40-year trip, you gave him a 7-digit number to dial on the spaceship's phone, he would only have one or two numbers dialed by the time he returned. Also, let us assume that we can watch the twin on the spaceship as he dials those numbers. It would take him years just to get his finger to the phone's dial, let alone dialing the numbers. And the clock on the spaceship would seem like it was stopped.

But let us realize the crucial point to be made here. During this whole event and even though time was "locally" distorted for each of you, neither you, nor your twin, jumped out of Now. It is simply that, your twin's atomic and subatomic interactions were slower than yours during this 40 years. This says that; your motion thru the space around you can affect the rate with which you measure time but time itself (now) is unaffected. An example of local time would be: if you were positioned on a quark and made measurements of time, distance and speed ((the exception being the constancy of the speed of light in a vacuum (empty space)), these measurements would only be relevant to observations made from the quark. The reason for this restriction lies in the fact that the quarks are in motion through the space within the protons, which are also in motion and these motions relating to the protons, are not exactly the same as the motions of the quarks within.

Hence, observers on the quarks and observers on the protons will have conflicting values relating to distance, time and speed ((the exception being the constancy of the speed of light in a vacuum (empty space)). This restriction is not confined to just quarks and protons, for quarks and protons are parts of atoms, which also have different motion and atoms are part of the earth which also has different motion, and so on, right up to the galaxies and all the other parts of the universe as a whole.

Recall, Dr. Einstein explained in relativity, simultaneity will only hold up within a local frame of reference (the embankment along the railway). But I have doubts about this. I say this with all these motions above in mind. Returning to Figure 17; His positions A and B, from where the lightning flashes originated, are located on the surface of the earth, and the earth is in motion, hence his frame of reference (A and B) is, in reality, in motion with the earth. And this means that any observer, at reference frame (A and B) will, in reference to relativity, already be somewhat shorter in length and his clock will already be running somewhat slow. (Fact 2)

From the above information, we must conclude, once the rays of light from the lightning flashes are set loose from positions A and B and with reference to the space around the rays themselves, position M is no longer motionless. The first part of the speed of light postulate says, the speed of light is independent of the motion of its source. But how about the space around the source?

Dr. Einstein said on page 329, in his book (Ideas and opinions), "The so-called special or restricted relativity theory is based on the fact that Maxwell's equations (and thus the law or propagation of light in empty space)...." Note that he always referenced the speed of light to empty space (in vacuo), but nothing else. With the exception of the speed of light, he used "point" frames of reference for his entire theory of relativity.

One simple example of this line of reasoning is; an observer on earth watches the moon travel by him and concludes, since he always sees the same side of the moon, that the moon does not revolve. According to Dr. Einstein's view (point frame of reference), he would be correct. But let us now use space as our frame of reference. An observer placed out in space would see the moon rotate on its axis, once every 28 days. Astronomers know this, but it seems that they fail to realize, the true frame of reference they are using, which is space itself. With other frames of reference all around.

What does this tell us? This allows us to say, if the speed of light is independent of the motion of the source (A and B), then the motion of the source (A and B) is independent of the speed of light. In other words, if X is independent of Y then Y must also be independent of X.

Follow here: Using Figure 1. again, along the embankment beside the railway we have positions A and B with position M halfway between (call these three positions reference frame X). Let us assume that reference frame X (located on earth) has a velocity v^2 in a direction through space toward position A (<=). Position M1 has a velocity v in a direction toward position B (=>).

An observer (call him O1) at position M with a clock (call it C1) assumes that he and positions A and B (frame X) are motionless. This leads observer O1 to believe, because position M1 is in motion at velocity v towards position B, relative to observer O1, that an observer at position M1 (call him O2) and his clock (call it C2) will experience a slowing in time.

This sets up a circumstance where Observer O1 at position M doesn't realize that he is in motion at a greater velocity through space, than is the observer O2 at position M1, and in the opposite direction.

Imagine observer O1's surprise when after a time Observer O2 returns to observer O1's position at M, and observer O1 discovers that observer O2's clock C2 has been running faster than his clock C1. This means, Observer O2's physical length actually increased and his clock C2 run faster, as he was put into motion.

I wonder why Dr. Einstein never addressed this particular consequence of his theory at this point in his writings? Could it be the fact that his calculations for the exact speed of light (186,000 mps) depends upon clocks, whose accuracies depend upon the earth's motion. In other words, if we assume that the earth is in motion through the space around it, we must also assume that all clocks on earth are in fact, running slow, due to the earth's motion. Hence the 1 second, we on earth use to measure the speed of light with, would necessarily be longer in time than 1 second for someone, in motion through space at a slower rate than us on earth.

As a result of this, the true speed of light, using space itself as reference, would have to be less than 186,000 mps. Example; Let us place an observer O2 with a clock C2 on a planet (frame 2) whose rate of orbital motion plus spin (v), is less than the earth's orbital motion plus spin (v^2) and then place an observer O1 with a clock C1 on earth (frame 1).

Using the view of astronomy, we see the earth (frame 1) orbiting the sun at a greater velocity than the other planet (frame 2). Therefore we would logically conclude that Observer O1 with clock C1 on earth at frame 1, is in motion through space at a faster rate than observer O2 with clock C2 at frame 2. Hence, observer O1's clock C1 on earth (frame 1) should run slower than observer O2's clock C2 at frame 2.

Let observer O1 on earth (frame 1), set loose a photon P1, for 1 second and then measure how far, through space, the photon P1 traveled during this 1 second with reference to his clock

C1. Next we let observer O2 at frame 2 set loose a photon P2, for 1 second and then measure how far, through space, the photon P2 traveled during this 1 second with reference to his clock C2. Now we compare the results found by both observers.

Observer O1 on earth (frame 1), after looking at the distance he measured, which was relative to his clock C1's measurement of 1 second of time (186,000 miles), found that the speed of light was 186,000 mps (d = rt). However, observer O2 at frame 2 after looking at the distance he measured which was relative to his clock C2's measurement of 1 second of time (180,000), found that the speed of light is 185,000 mps (d = rt).

Under relativity, the faster your rate of motion, the slower your clock will record time, hence observer O2's clock C2 ticked off its 1 second faster than observer O1's clock C1 ticked off its 1 second. This meant that the photon P2's distance traveled, relative to clock C2's measure of 1 second, was less than the distance photon P1 traveled, relative to clock C1's measure of 1 second. This of course, recognizes the constancy of the speed of light through space. Some might rebut here with the fact that distance can shrink with motion, but we're talking about distance through space here and not the length of shrinkable objects of mass.

We must remember that the motions for the earth, solar system, galaxy, local group, etc., would all play here. So what is the result of all this? Until we can calculate our exact velocity through the space around us, we must give our measures for the speed of light, only as an approximation and in reference to the earth only. Does it now become obvious why many within the scientific community have problems with the part of the law that states, the speed of light is independent of the motion of any observer? I will show a more detailed work, including the math, relating to this idea later within this work.

. It seems from what we have just discussed, that uncertainty is not just a component of Quantum Mechanics, but also a component of relativity. This caused someone to exclaim, "My kingdom, for a place to stand!"

As a side note: On page 30-31 in the book Einstein's Universe by author Nigel Calder, two physicists put calibrated cesium clocks on airliners and sent them around the world, one going east and one going west. The clock going east lost 59 nanoseconds when compared to a third identical cesium clock on the ground and the clock going west gained 273 nanoseconds when compared to the identical clock on the ground. The experiment was performed in order to see if less gravity (high altitude) and motion, relative to the rotation of the earth, affects clocks.

Notice that one of the airliner's clock actually lost time relative to the clock on the ground. This result does nothing in the way of concretely confirming my ideas put forth in this work, but if both airliner clocks had kept equal time, my work here, would have been badly damaged.

Have you noticed that there is always one property of nature that does not change; "The "assumed" constancy of the speed of light through space. This is nothing new, but just a note of future importance.

I see part of an equation here, relating to the constancy of the speed of light through space. I see an upper limit c, hence I see a lower limit, which would, by necessity, be zero. It is to be noted, Lorentz and Dr. Einstein, in dealing with time dilation, mainly addressed motion above zero and up to the speed of light. Dr. Einstein states on pg. 36 in his book Relativity-The Special and General Theory, "Of course this feature of the velocity c as a limiting velocity also clearly follows from the equations of the Lorentz transformation, for these become meaningless if we choose values of v greater than c."

The equations he speaks of here are Lorentz's first and fourth equations. Namely; t = 1 / √ of (1 minus v^2/c^2) seconds. In addition Dr. Einstein showed that, for the velocity v=c then √of (1 minus v^2/c^2) = 0. But he refrained, at this particular time, from showing the results of velocity v=0.

I can only assume that he believed, because space itself is empty (non-physical), all objects chosen as a frame of reference and not under acceleration, can be considered to be at rest, relative to themselves and this <u>empty</u> space around them.

While following the rules of mathematics, let us give the value for velocity, in the equation above, as zero. This would, according to the rules of mathematics, make the part of equation v^2/c^2, become undefined. Rather than start a mathematical argument by trying to use zero as velocity let us choose a velocity that approaches zero. By setting v = 1 x 10^{-10} meters per second we now have by t = 1 / √ of (1 minus v^2/c^2) "according to my calculator" = (1).

What this tells us is, by use of Lorentz's equations, "If an object were truly stationary, with reference to the space around it (S), its associated clock, which would now experience optimal molecular action, would measure time as "truly" t = 1 second." In other words, a clock sitting at absolute rest, with respect to the space around it, will keep optimal time, with reference to all the other parts of the universe in motion around it. Note that, not only am I saying this, but Lorentz's equations point to this also. In addition, the gamma factor table also allows for this. And relativity says this also if we really think about the factors for motion within it.

All motion, under relativity, is restricted to velocities that are less than the speed of light, which according to Dr. Einstein, objects of mass can never attain. But he is not completely clear as to why an object of maybe infinite mass cannot move through infinite space at speed c or higher.

The limiting factor, according to Dr. Einstein, Page 35-36, Section XII, is the velocity "itself", of the object in motion. He backs this conclusion with reference to the equations of the Lorentz transformation, which describe an objects length and time alterations, when that object is put into motion.

Quoting Dr. Einstein, page 35-36, "For the velocity v=c we should have the square root of I-v^2/ c^2=0, and for still greater velocities the square root becomes imaginary. From this we conclude that in the theory of relativity the velocity c plays the part of a limiting velocity, which can neither be reached nor exceeded by any real body. Of course this feature of the velocity c as a limiting velocity also clearly follows from the equations of the Lorentz transformation, for these become meaningless if we choose values of v greater than c."

This explanation sounds eloquent, but still leaves us wanting for a physical or mechanical definition. Looking at his idea simply, it says, an object's velocity (motion) changes measurements taken with clocks and rulers, with reference to that object, and the object's velocity has an upper limit of c. The reason velocity c cannot be attained by that object, is because an equation breaks down above velocity c. The question still remains; what "exact" physical forces or restrictions, keeps an object's velocity from reaching c?

We must start by looking at velocity itself in order to answer this question. What is velocity? Velocity, to put it simply, is a vector motion through space, and there is one ever present and automatic side effect of motion, which is distance. If you are in some motion (>0) then you are automatically traveling some distance (>0). So now we look at distance.

Let us place two points A and B separate from each other. By separating A and B we automatically create a distance between them. It can then be said, that point A occupies a position in space, different and separate from the position in space that B occupies. This says simply that A and B do not occupy the same point or place, but are in fact, some distance from each other. And this distance is always a "spatial" distance. I followed Euclid's examples here. This might seem elementary to some but I would ask for patience.

From the above, when we give an object some motion, we automatically give it some value of distance, relative to the space around it. As you can see, space is a necessity for motion and distance. So now we are led, by necessity, to look at space.

Space resides within and around all matter within the universe. Thus, we can say that space contains all matter and its motions. As an example: We think a cannon ball is solid but when we look within, we find molecules, with space within and around the molecules. When we look further we find atoms making up these molecules. The atoms we then find, are mainly composed of space with the exception of very tiny particles we call electrons, neutrons and protons. The electron was found to have internal gyroscopic properties (inertia) so we know that it has motion within and this tells us that space must also be within the electron. The neutrons and protons were found to have smaller particles within them also. So once again we find space within and around these particles.

Next, we bring the elementary particles (quarks and electrons) into contact with their anti-particles (annihilation) and the result is a huge outpouring of a still smaller particle/wave known as radiation (photons). This result, points to space as part of even what we consider to be, elementary particles.

To go still farther, we find that the photon is made up of many smaller particles known as quanta. Space is here also as you can see.

So what have we found? Molecules, atoms, electrons and nuclei all seem to be solid particles until we look inside them and find that they are actually just waves in motion through the space around them. This result holds all the way down to a single quantum. Here the story changes.

It is known that all matter contains energy. Quanta is seen as the smallest amount of energy that can exist. I have not read anywhere that a single quantum of energy has been separated into smaller parts. This says to us, quanta is the only true candidate that can qualify as a truly elementary (basic) particle. The fact that annihilation, does not go below the quantum level, tells me that quanta might in fact, be irreducible, at least with reference to any forces we know of (Conservation of Energy Law). I exclude the Big Bang here.

I will quote from page 43 in the book, ATOMIC PHYSICS by Max Born, "From the theoretical point of view the discovery of the positron is of the highest importance; for it confirms the theory of Dirac already mentioned, which on the intuitional side amounts to this, that neither electrons nor positrons are immutable elementary particles, but that by coming into collision they may annihilate each other, with the emission of energy in the form of light waves; and conversely, that a light wave of high energy can in certain circumstances become the source of a "pair" (electron + positron). There is experimental evidence for both processes."

What Born implies is; an electron (or any particle for that matter) that can be reduced into a still smaller particle (quanta in this case) cannot be considered to be a truly elementary particle. In addition, he shows that raw energy (quanta) in sufficient strength (ultra-λ-rays) can recombine in a temporary configuration of mass. Sounds like a Big Bang scenario doesn't it? Only smaller in scale.

Mix all this with the fact that these smallest particles (quanta) are generating a wave and the result becomes, all matter within the universe is actually particle/waves of energy in motion through the space around them. You can also say**,** all matter and even we humans, are basically quantum space waves. Look into a particle and you find waves. Look into the waves and you find particles. And so on and so on, right on down to single quanta. This is the particle/wave duality of nature around us! And space "physically" plays in all this. In fact, space is the ultimate particle (solid). All else is waves or configurations of waves.

If you have understood what I have just described you can now understand why the speed of light is the upper limit for all "mass" motion, in the universe. The universe itself consists of only three component factors, particle/waves of energy, space, and the pressure needed to produce the motions of these particle/waves of energy through this space. If mass is simply space waves (space's reaction to particle/waves of energy moving through it), and space is physical (shows resistance), then by the laws of motion, with reference to a medium, no object of mass can attain speed c (the speed of its quantum creator). In other words; if an object of mass were to attain a speed equal to that of light the object would lose its mass characteristics and in fact show its true identity which is light. It should be noted however, that space, internal to itself, possesses a reactive nature that is faster than the rate of motion for any waves traveling through it. I address this topic elsewhere within this work under the subject of Maxwell's electromagnetism.

This design also "mechanically" explains all forces and gravity. Gravity, under this design, is simply an inertial guiding path of least resistance (easiest direction to follow) due to the stretched configuration (expansion) of space around objects of mass. We can show the mechanical generation of gravity by setting up a grid-work for space or as I will do here, simply borrow Dr. Einstein's field and use it. But we must do a few alterations.

Dr. Einstein postulated that gravity is the warping of space-time. In my opinion he was correct, with the exception that we understand that the warping of time can be accomplished by two actions, motion and gravity. On the other hand, expanding space can only be accomplished by one action (energy in motion). Because, as we discussed elsewhere, if all matter is constructed simply by the action of energy in motion, then energy must also generate the gravitational fields.

Under this design, energy stretches the fabric of space as it moves through it. By use of the kinetic theory of gases, though under extreme pressure and density (fluid of great density), the particles within space itself react, as it transmits the energy through itself. This action excites (adds motion to) the particles making up the fabric of space. Hence, due to this motion, the mean free paths between the particles become greater (space expands).

This gives us a 3-dimensional space or field configuration with a positive curvature around objects of mass or photons. Dr. Einstein's 3-dimensional field, due to the gridlines curving inward toward the object of mass set within it, has negative curvature. Dr. Einstein seen gravity as mass contracting the "field" around it (negative curvature). I see gravity as energy expanding the "space" around it (positive curvature). Since mass and energy, within this work, are actually the same thing, I will not argue the fine points.

I do find it more than a little strange that with his negative curvature in relation to his physical field, Dr. Einstein failed to create an attractive force. It would seem, according to his gridlines, that his "physical" field would become more dense, the closer we get to the object of mass, hence, to me at least, this should cause a path of more resistance (repulsion) instead of a path of least resistance (attraction). This, I assume, is why the "hypothetical" graviton had to be put into play.

Within our physical space configuration, when a physical object comes through normal space and enters into a physical space that is stretched around an object of mass, this provides to that object, an inertial guiding path of least resistance. (See Figure 14.) As you can realize, under this configuration we don't need a graviton.

FIGURE 14.

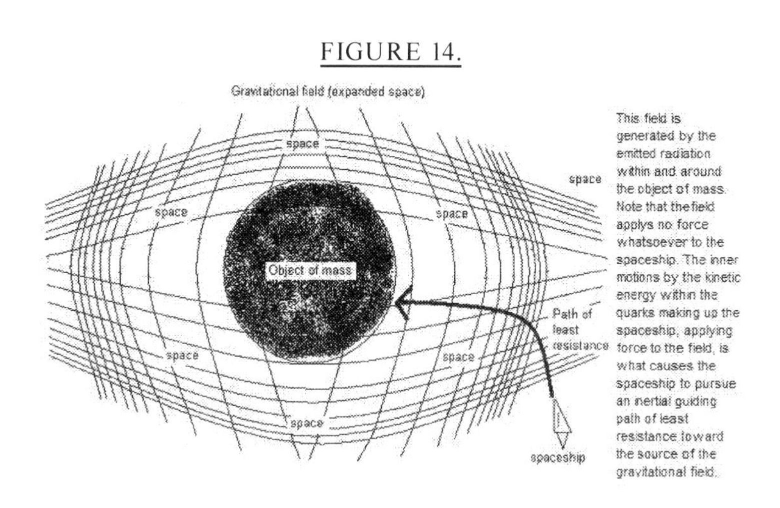

Excuse my art in (Figure 14.). If you will notice, space is stretched greatest at the center of the object of mass. This means that, as you go deeper below the surface of the earth (object of mass), in a direction toward the center of the earth, the affect of gravity continues to increase exponentially. Could this be why Newton had to reference the centers of the sun and planets in order for his equations to work? This also says that the rate of gravitational acceleration (32 ft/s/s) at the earth's surface, will increase the closer you go to the center of the earth. A simple experiment, perhaps in the gold mines of South Africa, would verify this idea.

As you view the figure above, keep in mind, we don't, as mass, move independently through the space around us. We are, in large part, the space around us. Just remember this, an object's gravitational potential, is realized by the effect its energy (internal motion) has on the space around it.

As we discussed above, once we understand that all forces and matter (mass and photons) and their motions are a direct result of quanta in motion, we then realize why the speed of light is unattainable by any object, with the exception of quanta itself. Put simply, "All matter in the universe is made of energy in motion at c with the exception of space itself, which transmits, and governs the speed, of the energy." And what is space itself made of? I hinted at this question before. The answer can be arrived at simply by bringing a single particle/wave of quanta to rest.

With the understanding that I view everything within the universe as containing "some form of physical matter", even to include energy and space itself, I define the universe as an infinite and perpetual, transformer or engine if you will". The laws of thermodynamics forced me to this conclusion.

Thermodynamics tells us that all mass matter will eventually breakdown (entropy). We discussed this before using an electron. But breakdown into what form? The answer can only be energy, which can be neither created nor destroyed. Professor George Gamow showed that "this part of the cycle we're in now", started with" kinetic" energy (Big Bang). I believe after all is said and done (entropy), this part of the cycle will end with "dormant" energy in space itself.

The ultimate aim of this work is to show that space is not empty at all, but is in fact made up of a physical substance (particles of energy brought to rest). Black hole theorists, by use of the uncertainty principle, borrow "virtual particles" from this supposedly <u>empty</u> space regularly. This allows for a hypothesis to be put forth whereby, for a short period of time (less than Planck time) matter can be drawn from and returned to, space itself.

With this in mind, if space contains particles of any type, then these particles must contain energy, hence, space itself must contain energy. And how would space acquire energy? The simplest answer becomes, by absorbing quanta.

Out beyond all gravitational fields connected to all the energy ejected by the Big Bang, the fabric of space is far denser than it is locally. Light traveling out into this super-dense region of space is shredded from the photons, absorbed and brought to rest. This points to a conclusion; not only is mass and energy convertible but so to is space. To rephrase; Mass is energy in non-linear motion (mutual orbits within quarks). Photons are energy in linear motion. Space is energy at rest.

The question now becomes, what would a particle of light look like when brought to rest in space. A single particle of light, before it is absorbed by space, by my calculations, would have a diameter of roughly 10^{-24} cm. Note here that the particle of light is contracted and has very little internal motion, due to its motion through space at speed c. Once this particle of light is absorbed by space and brought to rest, it would expand due to the rate of motion of the particles and waves within it increasing and would now have a diameter of approx. 10^{-23} cm. Note here that the particle of light is smaller in diameter than a particle of space but do not be led to believe there is no sub-space. I leave that idea to some future aspiring science student.

Again; note that the diameter of a single particle of light is less than the diameter of a single particle of space. This explains how a physical space allows physical light to move through it. The Lorentz contraction makes this idea possible. This configuration is also a pressure producing factor and I will talk of this later.

What does all this say? It says that, the ultimate conclusion that can be drawn from the atomic theory is, even the particle/waves of energy contain even smaller particle/waves of energy. At this point we are within what I call sub-space. Particle diameters here would be on the order of 10^{-34} cm or maybe smaller. I believe pressure knows no bounds!

Some might question how we can do credible work at this level so I will now address this. I would ask that anyone who reads the following, take note of the result of the work and not the details themselves.

Space is where the electromagnetic field plays. And guess the speed with which it plays? That's right, the speed of light as Maxwell showed. On pg.74 in the book, The Elegant Universe by Brian Greene, he says, "Einstein was able to calculate how fast disturbances to the fabric of the universe travel and he found that they travel at precisely the speed of light".

Dr. Einstein's calculations here were in relation to his work on gravity but as I see it, he unwittingly, has arrived back at Maxwell's physical space (electromagnetic field). If Dr. Einstein's universal fabric (the field) can feel and transmit the effect of gravity then I find it more than possible, that this fabric also feels and transmit the effects of electromagnetism, and at the same speed.

This says that the speed with which electromagnetism propagates through space, has nothing to do with the speed with which it transmits light, but in fact, has more to do with the reaction time of Dr. Einstein's field fabric (space itself) as we discussed before.

It is known that the source for an electromagnetic field is electrons put into motion. But can space itself feel this? The answer, under a configuration that contains a physical space, is definitely yes.

Space, as we discussed before is extremely dense with the matter making it up having mean free paths equaling zero between the very small particles making it up. Hence, if an electron is put into motion within it, space will feel and react to this motion, according to Maxwell, at a speed equal to the speed of light. Hence, the electromagnetic field is propagated within space itself and not thru space. And the rate of propagation is the same as the speed of light, and also the same as Dr. Einstein theorized for gravity.

With all this said, the question becomes, how do we re-activate energy after it has come to rest in space? The answer is simple, "pressure". Space, under pressure, erupts, which gives us a huge outpouring of radiation (Gamow). The force (released space pressure) with which the Big Bang occurred, causes the space around this event to become greatly expanded due to all the energy radiation being expelled from space. This also causes the radiation to accelerate to a speed in excess of the normal speed of light (tachyons). The first huge waves of tachyons eventually travel out into the space around this event, where space is less expanded and they slow down to speed close to c as a result. Note; with the exception of ever-present pressure, the only force present at this stage in the Big Bang is radiation pressure.

The following waves of tachyons now overrun the first waves and this causes a piling effect (mass creation) and "quantum gravity" is born here. Simple atomic structures where first to be formed (mostly hydrogen and helium but possibly all the way up the periodic table to some carbon atoms. Classical gravity meanwhile, is working to gather as much of these basic atoms together as it can (stars, planets, moons, and galaxies).

Hence we end up with great clumps of basic matter (stars) which under the force of their own gravitational and electric forces, build more complex atoms. Next, after tens of billions of years, all this mass eventually decays back to radiation (this is entropy). Lastly, as we discussed above, space reabsorbs the single quanta of radiation as they eventually travel out into non-gravitational space. Let me reiterate**,** "The universe exists in a configuration such that, the entropy it must undergo, is the very cause for its rejuvenation".

Looking at the universe as an "infinite and perpetual" entity, with its high temperature reservoir being the energy/mass within the universe around us, and the low temperature reservoir being space itself, then we must, according to thermodynamic laws, assume, at some time in the far future, both the high and low temperatures will even out and the universe will, in effect stop. But this has never been the case since we know the universe has already existed forever (see Axiom #1) and is still doing work now.

This is because, under the present arrangement, the energy (motion) connected to the particles of light, once brought to rest when space eventually absorbs the quantum particles of light, is converted to pressure within space itself. According to the second law of thermodynamics, the natural direction of heat flow is from a reservoir of internal energy at a high temperature (light) to a reservoir of internal energy at a low temperature (space), regardless of the total energy of each reservoir.

Within this work, the universe is "spatially" infinite in size but also made of a transformable physical substance (expanded particles of light at rest) and contracted particles of light in motion at c. Because the universe is infinite, having no outer edge, energy cannot escape, hence it can be considered as a closed system. In addition, because this system is totally physical, pressure can be applied to it as a whole (Pascal's principle).

But all this by itself, still might not be adequate in meeting the requirements needed in order to show the "perpetual" nature I ascribe. Thermodynamics shows, after all pressure within a closed system is spread out equally, work done by the system, stops. This was the point I arrived at when I had to consult with GOD. This is when I came up with a configuration whereby the universe contained more substance than it could properly hold. (See perpetual motion on pg. 11)

You might ask why I would want to show the universe as "perpetual" in nature? The 1rst law of thermodynamics tells us we can't get something from nothing. Then how did we, or the universe, get here and how can the universe still be doing work? With reference to time, it must be understood; when analyzing and combining the principles of the 1rst and 2nd laws of thermodynamics, we and a working universe, are an impossibility. Yet here we are, and working too! Now to recap.

First, the universe never started. It simply transforms itself using pressure to do the work. This meets the requirements connected with the 1rst law of thermodynamics.

Second, the universe contains more substance than it can properly hold. Only with some of the particles and waves (substance) within it, in motion and contracted, can the universe contain all the substance within it.

Example; if you take a cup and fill it to the brim with a substance, you might think the cup is, in fact, full and no more substance can be added. But if you then stir some of the substance, within the cup, at speed c, the substance within the cup will contract and you can then add more substance. Now imagine doing this operation to the infinite universe as a whole.

Third, when the motion (energy) within the high temperature reservoir system such as this, is decreased (light being absorbed and brought to rest by space where it then expands), the pressure (work) on the low temperature reservoir increases. Here we apply Pascal's principle which says, an external pressure exerted on a fluid is transmitted uniformly throughout the volume of the fluid.

This action causes the pressure upon the particles making up space itself, to become so great that the mean-free paths between the particles making up space itself, go into annihilation with each other (Big Bang). Understand, the universe being physically and spatially infinite means, pressure cannot cause it to "explode" outward, so it reacts internally (implodes) instead.

Fourth, Space implodes (Big Bang) from space's internal pressure, which causes the particles making up space to be accelerated out into space again, contract, generate a space-wave around them and become what we know as energy. Under this configuration, some of the total pressurized static matter within the system is converted (Big Bang) into motion (kinetic energy), hence the total system pressure decreases as its internal motion (kinetic energy) increases. And as the internal motion again decreases within the system (entropy), total system pressure increases again.

The 2nd law of thermodynamics tells us that a "perpetual motion" engine cannot be constructed and I agree. However, I believe one does exist and the universe itself is my candidate. The infinite universe is a continuous (perpetual) engine or transformer if you will.

I and the 1rst law of thermodynamics could not allow the universe to have a beginning to its existence (something for nothing). In addition, the universe defeats the 2nd law of thermodynamics by using its end product (entropy) to do the work needed to rejuvenate the system. I hope it is noted that only with a physical space could something like this be possible. Only with a completely physical universe could I put the extreme pressure upon it, needed in order to make it do perpetual work. This says simply, "The universe itself abhors a void".

In defense of my idea of a physical space I will say, Theoretical Physicists borrow from it. Dr. Einstein curves it. Light which is transmitted by it generates a frontal kinetic wave in it while stretching it. Mass spins it. Electrons drive it. Electromagnetism is propagated by it. Clocks slow down in it and objects foreshorten in it. To put it bluntly, "Space is the parent substance for the universe around us and it is under pressure. Hence, the universe is under pressure and pressure makes it work". Under this configuration we see; all forces within the universe are really just manifestations of the one true force, which is pressure. But most of all: We, as objects of mass, don't just move thru the fabric making up space; we in fact, use the fabric of space we travel "within" for our existence itself. This fact shows that we, as objects of mass, basically exist as a combination between the fabric of space we travel within and the kinetic energy within us. Understand: we physically use the space we travel within. Note: I can point this out but it's up to you the reader to make the leap.

Particles such as electrons and protons are simply reactive space waves, which are generated by particle/waves (quanta) in motion within them. This would mean that we humans are a product of wave motion which is directly connected to energy in motion. We might see ourselves as motionless (at rest) in reference to the universe around us, but in fact we are always in motion at the speed of light. This is due to the fact that we contain energy, and this contained energy is in motion at speed c. In other words we are simply a spatial effect of the energy in motion within us. And this points to a conclusion that says, "we humans and all other objects of mass are in reality just effects of energy in motion". And because this effect is physical (real), therefore we are real, though temporarily.

And under this configuration, the universe is simply space and time (energy at rest and energy in motion), with pressure being the catalyst for change. This says that time, for us, within the available energy reservoir, moves at the speed of light. It might seem that I hammer way too much on certain ideas but that's the only way I know, to advance my ideas. With that said, let us continue.

Most believe that all the physical properties within particles, convert to energy during annihilation. I believe energy is never converted, except when reabsorbed by space. This says that if quarks, electrons, and neutrinos have no nucleus then they are composed solely by particle/waves of energy (quanta) traveling in orbits within them at speed c. And space itself transmits and governs the speed c with which energy (quanta) moves through it.

The scientific community at present, believe that light slows down while traveling through a medium. Let us place a piece of glass (medium) out in space and send a single quantum of energy through it. It is believed that the quanta enters the glass, slows down, and "instantaneously" reaccelerates upon leaving the other side. But we must remember, from the discussion above, that the medium itself (glass) is simply "spatial" quantum matter waves as a result of kinetic energy. The quanta does not slow down. It's simply that it takes longer for the quanta to traverse the forces set up by the matter waves within the glass.

In other words, light does not move through a medium in a straight line. Its path is dictated by the forces it encounters within the medium. You might also say, light undergoes many reflective changes to its path as it travels through a medium, hence it takes more time for the quanta to traverse the medium. I will postulate here, "Changes to the speed of light can only be attained by manipulating the density of the space which transmits the light."

This says, the speed of light should be slightly faster out in space than it is here on earth. The difference in the strength of the gravitational fields here on earth and out in space can be used as a direct substitute in solving the equation connected to this idea.

To put it simply, out in space where the fabric of space is stretched less (less gravity), the faster, quanta is transmitted through it. Close to an object of mass, the more expanded the fabric of space becomes (more gravity), and the slower it will transmit light through it. Hence, because the speed of the quanta within the matter making up a clock, when placed out in space, will increase, then so to should the clock's molecular action. The clock will run faster than it would here on earth.

Dr. Einstein postulated this end result also, but connected it to only an effect of gravity. But as you can see, the speed of energy through the space around it, within the mass making up the clock, is partly responsible for this result.

Anyone familiar with Dr. Einstein's work, relating to gravity, will immediately recognize that this result above agrees with his work (see Nigel Calder's book, Dr. Einstein's Universe, pg. 30-31. The difference is that he uses a virtual field within space while I use physical space itself. As a pun I will say, "We all need our space".

As an aside, if we assume that the generation of Dr. Einstein's gravitational field, is the result of gravitons being emitted outward into the space surrounding them, then we must insert an exception into the laws of motion. Newton's 3rd law says that the result of any action will be an equal and opposite reaction. From this, it becomes clear that attraction, in a direction toward the applied force (gravitons), defies Newton's 3rd law.

In addition, we must now ask, what happens to a graviton once it collides with an object?

Is it absorbed, is it reflected, does it disappear? What happens when two gravitons meet? Do they then join, or do they annihilate each other, and wouldn't actions such as these be detectable? Finally, how can a graviton escape a Black Hole when not even light is allowed to escape?

I read somewhere, "If an idea is illogical (challenges empirical data), then it has better than a 90% chance of being wrong." I would be obtuse to believe that my work here could not also fall into this category. That is why I continually try to reference my results to experimentally proven ideas and laws.

Several different theories have been put forth as to the beginning of the universe. A Steady-State theory and The Big Bang theory are the two most popular theories at present with the Big Bang theory being the one most adhere to, by the scientific community. This Big Bang theory sees both space and time as having a beginning.

The major problem with this part of the theory is not with time, as you might guess, but with a beginning or creation of space. If space is seen as empty or void of matter (nothing), then how does one theorize its creation? In addition, how does one then bound this nothing? Do we now have two types of nothing? I have to believe that man can find the right answer to any dilemma, if he probes and ponders long enough.

You might now ask how this, effect of space I speak of, can be shown. I will begin by describing the strange results that occur when an object is put into motion through the space around it. The equations used for the Lorentz transformation point to a conclusion that, when an object of mass is put into motion through the space around it, the object becomes shorter in length and its internal clock is slowed or dilated. In addition, the faster this object moves through the space around it, the shorter in length it becomes and the slower its clock runs.

But how does the object of mass know it's in motion and how does it know the exact magnitude of its motion? Unless Maxwell's demon is playing here, there can be only one logical answer. The internal parts of the object of mass, when this object of mass is in motion through the space around it, feels a resistance or effect as the space passes through it, or as it passes through the space, as you like. This says, "Space presents a physical resistance or force, to any object in motion through it." The exception to this is light, where space both, accelerates and resists it.

As we discussed before, if the object of mass were truly at rest relative to space, with reference to Lorentz's equations, it would show its ultimate length and its clock would be running at its optimum rate. These results are, just like Dr. Einstein's, a direct consequence of the equations of the Lorentz transformation. I expanded the results of the equations, by reducing the velocity of the object, in reference to space, to zero. You might also say that I reduced the gamma factor to its lowest limit. Dr. Einstein, due to his belief that space itself was empty, had no reason to pursue this particular line of thought.

I built this assumption by imagining the universe somewhat like a huge pool table with the balls interacting all over the place. I have noticed that sometimes a ball, after interacting with several other balls will end up motionless with respect to the other balls and the table. I assume that the universe should have more than a few of its balls (objects) close to being motionless or in fact motionless, in space. Chunks of matter being scattered at different rates and directions, during Super Novas, makes this idea more than just plausible.

In my opinion, the reason Dr. Einstein could not advance a sound mechanical reason for why objects of mass were not allowed to attain the speed of light, is because he seen space as empty or nothing if you will. He postulated that an object becomes more massive as its velocity increases because the velocity itself (energy of motion) adds mass to the object. And the more mass or inertia an object has, the harder it is to accelerate the object. But now we must ask, where is all this extra mass coming from?

On page 328 in Dr. Einstein's book – Ideas and Opinions, he references J. J. Thompson who pointed out that an electrically charged body (we'll assume an electron here) in motion must, according to Maxwell's theory, possess a magnetic field whose energy acted precisely as does an increase of kinetic energy to the body (electron). If, then, a part of kinetic energy consists of field energy, might that not then be true of the whole of the kinetic energy?

What Thompson is saying here is, the increase of the electron's kinetic energy seems to come from the field or in our case space itself. Let us analyze this.

If we place an electron far out in space and accelerate it to 99.999% c, the electron's mass/energy, in Dr. Einstein's view, is greatly increased. But we must remember, all we have to factor with here, is the electron and the electron's velocity, with respect to the space around the electron. This configuration, mathematically, and with reference to the conservation of energy and mass, will only work with the increase of mass/energy, being an effect caused by the electron's motion through the space around it. And we have seen through High Energy physics, this effect is real.

So what does this tell us? With only an electron and its motion through the space around it to factor with, the effect of extra mass/energy, must come from space itself, just as Thompson theorized above. The motion of the electron, through the space around it, produces this effect. And that motion is governed by space itself. In my estimation, this is the only mathematical configuration that preserves the conservation of mass and energy relating to the above action.

As an aside, if we recall Newton's first law of motion which says, "A body in motion remains in motion (with constant velocity) unless acted upon by an external force", but note; when we apply this law, to what we discussed above, we find a problem.

Follow here; Let us place an object out in space beyond all gravitational fields and give it a constant velocity v. Dr. Einstein's postulate says that this body will have an amount of mass greater than it would if it were at rest. But thermodynamics says that any increase in a system's energy comes at a cost. The cost for this extra energy is, as we know, the velocity v of the object.

But this cost must be paid back for thermodynamics also says that, no closed system can generate as much energy as it consumes (perpetual motion). If we see this body, which is in motion at velocity v, as a closed system (alone in space), and generating extra energy (energy of motion), then by the laws of thermodynamics, it must eventually slow down and come to rest relative to the space around it, and give back its extra energy of motion. Either Newton is mistaken or the laws of thermodynamics are wrong. It's my guess that thermodynamics will govern here. I doubt that Newton was well versed in what we call entropy.

In addition, I find that under deep analysis, Newton's 1rst and 3rd laws conflict with each other. His first law says, "A body in motion remains in motion (with constant velocity) unless acted upon by an external force". But wouldn't a body in motion through space qualify as an action? With this in mind, let us apply his 3rd law (reaction) to this action of a body in motion and the resulting energy of motion it is generating.

This says that the energy of motion is generated as a reaction, to the action, of a body in motion. In other words, the object is generating extra energy due to its motion. And as we discussed above, thermodynamics has to be a factor into any such energy manipulations.

We are now led to ask, with all this shown, what would this effect look like? It would look similar to DeBroglie's pilot waves. And Schrodinger's wave mechanics show some of the effect I speak of above. Schrodinger evidently, could not imagine a frontal kinetic (pilot) wave such as an electron has. The reaction by space, to an electron moving through it would be the generation of a frontal kinetic space wave, which would then generate a long harmonic (standing) space wave behind. This is why experimentalists have found, through empirical effects and statistical analysis, the electron resembles a smeared out wave, orbiting the nucleus.

The electron drives the space ahead of it to form the frontal kinetic wave. The reaction, to this action, by space, is to reform itself after the kinetic wave and particle (electron) has passed through it. This reforming creates a vibrational space wave which is analogous to a guitar string vibrating until it stops. Thus a standing wave is produced as an electron moves through the space around it. Remember, all waves and all particles, with the exception of kinetic particles of light, are simply "space waves". This explains why the number of electrons, allowed within any one orbit surrounding the nucleus, is limited (Pauli's exclusion principle).

To continue; we must realize that this frame of reference (space) is not a point-frame of reference like it would be if we chose a frame of reference on earth, like the tip of the largest, of the three pyramids in Egypt. This frame of reference space, is an all encompassing frame of reference. It resides within and around every piece of matter in the universe. It is also presently seen as empty, yet somehow, as we have seen, restricts the motion of all objects, up to and including light quanta, that move within it.

We, as physicists, must try to measure and hence, explain all of nature around us. That's our job, so to speak. As discussed before, basically we must have three things in order to accomplish our mission, a ruler, a clock, and a reliable frame of reference. The use of the ruler and the clock seems simple enough, that is until you start to factor in rates of motion, especially when related to different frames of reference with different motions for each frame of reference (relativity). The motion, it has been shown, can cause the results found by use of the ruler and the clock, to become vague or misleading.

We still wrestle with this dilemma today, even with all the hi-tech gadgets and theories at our disposal. What is needed is a point-frame of reference that can be considered as stable or "at rest", when compared to all other frames of reference and their individual motions. So let us hypothetically create, a universal frame of reference, while staying within the boundaries set up by the laws of physics.

I will call this universal frame of reference, imaginary point **(^)**. Mathematicians have brought into use several imaginary numbers (one of them being the square root of minus-one), to assist them in they're work. History has shown that they have had great success with them. I will follow their lead here. Let us now travel out in space to the location where the Big Bang occurred.

I would ask the reader to bear with me here and keep an open mind as to this imaginary point **(^)**. Its location is out in space away from all gravitational fields. Its relative motion through the space around it equals zero. It is the one point of reference that will be relative, to all motion (clocks) within the universe around it (Hubble bubble). A clock placed at this imaginary frame **(^)** would run the fastest and truest a clock can possibly run. Under Relativity and with reference to Lorentz's equations, this clock would be the standard for all other clocks.

Now, let us now create a co-ordinate frame of reference, relative to our imaginary frame **(^)**, under the rules of the Cartesian system of co-ordinates. The first problem we must deal with, relates to finding a way to bring our point-frame of reference (call it S) to rest, relative to the imaginary frame **(^)**. This means that we must stop our frame S's motion, relative to frame **(^)**, which is at rest with the space around it.

What I mean here is, we must make sure our frame S is at rest, in relation to all directions of the space around frame S. Realistically, at some point in the future (3000 AD), given our present rate of technological advancement, this process might be achieved in just the way I'm getting ready to describe.

Because I'm working in three axis I will need three space ships (call them N1, N2 and N3, with plenty of fuel. I move frame S out into space and put an observer (call him O1) at frame S with a clock (call it C1). Next I establish three straight lines x, y and z, 186,000 miles in length, perpendicular to each other and intersecting, at they're middles, at frame S. I then have ship N1 accelerate out and back on axis x, going faster all the while. I do the same for ship N2 on axis y and finally I do the same for ship N3 on axis z. The ships have plenty of fuel and can continue accelerating for as long as needed. (See Figure 15.)

FIGURE 15.

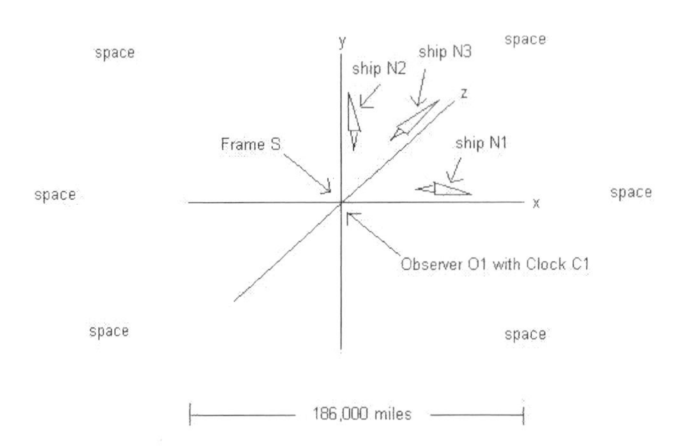

I now have ships N1, N2 and N3 continuously accelerating past observer O1 on paths x, y, and z, of 186,000 mile lengths. Observer O1 sits at the intersection (middle) of these 186,000 mile paths with his clock C1 and he starts timing each ship as they pass by.

Observer O1 notices that, after some time has passed, ship N1's lap-time, according to his clock C1, has barely changed even though ship N1 is continuing to apply acceleration (fuel). This means that ship N1 on axis x has reached a velocity close to the speed of light, for even though ship N1 continues to accelerate, his lap-time is barely changing.

So now observer O1 checks his clock C1 and because he knows the lap-distance (186,000 miles), he calculates ship N1's speed, relative to his frame S. It turns out to be 80% of the speed of light (148,000 mps). But observer O1 also knows, by the acceleration factors, that Ship N1's actual speed, in reference to space, is actually approaching 186,000 mps (- .001%). Why do we have a 20% discrepancy?

Observer O1 realizes that his frame S is also in motion in some direction through space. Using the results above, observer O1 then calculates that, with reference to the direction through space of the x axis, his frame S is in motion at approximately 20% of the speed of light. Observer O1 now puts his frame S and its axes x, y and z, in motion at 20% the speed of light, along the x axis, either left or right until his clock agrees that ship N1 is in motion at near the speed of light (-.001%). Observer O1 then has all three ships adjust their paths to match the revision he made to frame S and continues.

Observer O1 does this same exercise with ships N2 and N3 along their respective axes and keeps rerunning the full exercise with all three ships until observer O1, using his clock C1, has verified that his frame of reference S is stationary relative to all three ship's motions (99.9% of c) along their respective axes.

What we have done here is to use the restrictions, inherent to the Constancy of the Speed of Light Law and also to Lorentz's equations, to find us a stable place, somewhere in space, at which to stand and measure. This frame S will also be a direct co-ordinate to our imaginary frame of reference **(^)**. Note that I act here under Newton's assumption of a physical space. And what exactly will having this frame S, which will be equal in all respects, except for location, to imaginary frame **(^)**, do for our measurements taken at frame S.

To begin, we will have what I call <u>universal standard time</u>. A clock C1 placed at frame S, and because it is at rest with respect to space itself, will run as fast and true as a clock can "naturally" run. With an object's velocity at zero and with reference to frame S, the equations of the Lorentz transformation will not apply. Only when the object is put into motion, with reference to frame S, does the transformation take effect.

Also, if we place a ruler at frame S it will have the longest measure of distance naturally allowed. In other words, by being motionless relative to frame S, the ruler will not experience Lorentz's foreshortening. A ruler placed at frame S will truly measure 1cm = 1cm, and a clock placed at frame S will truly measure 1 second = 1 second. This is because motion (Lorentz's equations) will not be a factor.

It should be evident that this frame S can exist. As we discussed before, some clumps of matter ejected from a super nova should fit this bill. The equations of the Lorentz transformation point to it as a necessity. Velocity within the equations for the Lorentz transformation has an upper limit c, which says that, with velocities above this upper limit, the equations of the Lorentz transformation do not apply. This points to a lower limit for velocity, whereby the equations of the Lorentz transformation do not apply either. This says that, "Velocities c and zero, are the two limiting factors for Lorentz's transformations." We will call this (Fact 4). At velocity $v = c$, the result of the equation is 0. At velocity $v = 0$ the result of the equation is 1.

Many within the scientific community believe that Dr. Einstein's Relativity points to a conclusion that says, there cannot exist a true and stable frame of reference where one can stand and measure the complete universe around him. I see the opposite. Let us look at the Facts 1, 2, 3, and 4.

Fact 1 says, "If it takes the light ray more time to catch you when you are in a motion ahead of the advancing ray, then it also takes the light ray more time to pass you."

Fact 2 says, "The theory of relativity says, when an object is put into motion, its physical length is shortened and its internal clock will run slower."

Fact 3 says, "The constancy of the speed of light through space."

Fact 4 says, "Velocities c and zero, are the two limiting factors for Lorentz's transformations."

It can only be evident that these four facts all have one common frame of reference which is space. It might seem philosophical to say that everything happens in the space around it, but this is not only a philosophical truth, but also a physical truth.

Relativity says, there is more than one way to look at any event and all can be equally true depending upon the frame of reference you use. But with Frame S as the frame of reference this cannot be the case. An example: Using Frame (S) as the frame of reference, let us return to an earlier discussion within this work.

If a spaceship passes an observer O1 on earth at near the speed of light the observer O1 can say, with reference to himself, "The spaceship passed me at near the speed of light." On the other hand, an observer O2 on the spaceship can say, with reference to himself, "The observer O1 and earth passed me at near the speed of light."

According to Relativity, observers O1 and O2 would both be making a true statement. But as we discussed above, space (frame S) knows the difference. And because space knows the difference, and space applies a differential resistance to all objects (atoms and molecules) in motion through it, this means the atoms and molecules also know the difference. The Laws of Chemistry and Biology can be used to back me on this point.

Let us assume that the spaceship with observer O2 continued in motion at near the speed of light, in reference to observer O1 on earth, for one full month and then rejoined observer O1 at his position. Let us also note that neither observer, O1 or O2, shaved during this period of time. The Laws of Biology will show that, with reference to observer O1 on earth, going one month without shaving will produce a beard of considerable size.

This deduction, connected with the Laws of Biology is then found to be accurate for Observer O1, who has a beard several inches in length, but observer O2 still looks clean shaven. What does this result tell us? The rate of molecular action, with reference to observer O2 on the space ship, has been slowed when compared to the molecular action of observer O1 on earth. And this takes us to the Lorentz transformation (Fact 2) which can explain this difference in length of beards.

Fact 2 says, "The theory of relativity says that, when an object is put into motion, its physical length is shortened and its internal clock will run slower." If we assume, from the above, that Observer O2's molecular action is slow (he has no beard), relative to observer O1's molecular action, then we have the answer and the answer only applies to a single configuration that has observer O2 in motion "within space", at a faster rate than observer O1.

Please realize the importance of what I have just showed you. The effects of the force connected with acceleration is not the same as the effects of the force connected with a gravitational field. Acceleration is the result of an applied, outside force, while gravitational force is the result of the particle/waves (quanta) within us, taking an inertial guiding path of least resistance within the space around us. To put it differently, acceleration is the result of a foreign force being applied to us, while a gravitational force is the result of us applying our own force (energy in motion within the quarks within us) to the asymmetrically expanded fabric making up space. Acceleration is an active force while gravity is a passive force, you might say (inertial guiding path of least resistance). This says, "We might not know the difference between acceleration and gravity, but the atoms within us, and space itself, does".

I hope that we understand another result obtained from our Spaceship experiment above. The Laws of Chemistry (molecular action) pointed directly to the necessary existence of imaginary frame (^). Molecular action, as we have seen above, knows by comparative effect, who was in motion at the faster rate, within space.

As we mentioned before, Dr. Einstein postulated that the faster an objects motion, the greater becomes its mass. From what we discussed before we find that we need to change the word mass, in relation to an object in motion, to read "energy of motion". When an electron is motionless with respect to space (frame S), its internal energy (mass/inertia) is at its greatest potential and its external energy (energy of motion relating to the electron as a whole) is at its least potential. When an electron is in motion at near the speed of light its internal mass is near its least potential while its external energy (energy of motion) is near its greatest potential. This again shows why an object, once it is put into motion, will contract and its clock will slow down.

Lorentz shows that an object shrinks as it is accelerated. Along with this we know from Maxwell's work that an electromagnetic field is real or physical (an amount of electromagnetic force per area of the field). An electron put under acceleration will shrink in size (Lorentz transformation). Hence it becomes more dense per its size and with greater momentum, will penetrate the fabric of a physical magnetic field more easily. It only seems more massive when in fact, it is simply smaller and has more momentum.

Again, the electron looks like it has more mass but in reality, it is smaller and more compact and has momentum due to its velocity. Take this together with the fact, the structural fabric, making up the magnetic field is still the same density per area, and we can now see why the electron seems more massive as it passes through the field.

In addition, recall that the electron generates a frontal kinetic wave (DeBroglie) as it is put into motion through space. This is the direct physical effect of the energy of motion connected to the electron in motion.

Also, if Lorentz's equations are to reflect reality then they must reflect reality at both, the microscopic levels and the macroscopic levels, in nature around us. The equations of the Lorentz transformation say, as an object (electron) is put into motion, its size is decreased and its atomic action slows. I find it hard to accept that an electron's mass will increase after these internal "restrictive" motional effects are applied to the electron.

As we discussed before, if we assume that an object's mass energy is optimal when that object is at rest relative to the space around it, then we must also assume that this objects mass energy becomes minimal the closer to the speed of light its velocity becomes. This says, if an object of mass where to actually achieve the speed of light, all you would see is a photon where the object once was.

Recall what was said before; that the electron is simply particle/waves of light traveling in orbits (quark). If particle waves of light can travel through space at only one speed and the electron was to somehow attain this same speed, then the particle/waves of light within the electron would be unable to travel in the orbits, but could only travel in the same line as the object of mass was traveling, hence the object of mass would now be a photon. In addition, mass would now show its true face which is energy in orbital motion.

I'm not sure that hi-energy physicists realize it but one of the main reasons they cannot accelerate an object of mass to the speed of light is because the velocity of the force they are using to propel the object is only equal to the speed of light. Now add this with the fact, particles are known to squiggle (non-linear motion) due to their internal gyre (inertia), as they traverse an electromagnetic field. This action presents, what could be called, a resistance to directional accelerative motion. It would be like trying to herd cats. Now let us look at light itself.

Fact 3 relates to "The constancy of the speed of light through space." On page 339 in the book Ideas and Opinions, by Albert Einstein, he postulates, by use of the equation $E=mc^2$, mass and energy are equivalent. E is the energy contained in a stationary body, m is its mass and c is the speed of light. The important part of the idea expressed here is that, E is the energy <u>contained</u> in a stationary body. We will take this fact literally and we will call this (Fact 5).

Taken literally, Fact 5 says, a body (we'll call it an electron), contains energy, better known as quanta. Let us now "arbitrarily" apply Fact 3 to Fact 5., which now says, "not only do bodies of mass contain energy but the energy contained within these bodies are always in motion at speed c". **<u>This is the key to the quantum world</u>**. The scientific community at present believes that mass has to undergo a transformation in order to become light. They also believe that the resultant light attains its speed instantaneously. Any knowledgeable reader that has been paying attention to what I've been showing will know that these beliefs are unfounded.

I have questioned many learned people within the scientific community and almost all of them have a problem with the idea that time stands still while an emitted particle of light accelerates up to its ultimate speed c. As noted before, the equation $d = rt$ is insufficient in this circumstance. What Fact 5 now tells us is, "The electron's mass is generated by the action of quanta running in circles at velocity c within the electron. This approach also explains a problem in Quantum Mechanics that has existed for years with no reasonable explanation. Namely; why do particles exhibit gyroscopic properties even though they, as a whole particle, are spinning very little? (See Figure 16.)

But now the question becomes, how can all these particle/waves of light run in such small circles and collect together in such a small space, to create an electron. The answer is quantum gravity.

FIGURE 16.

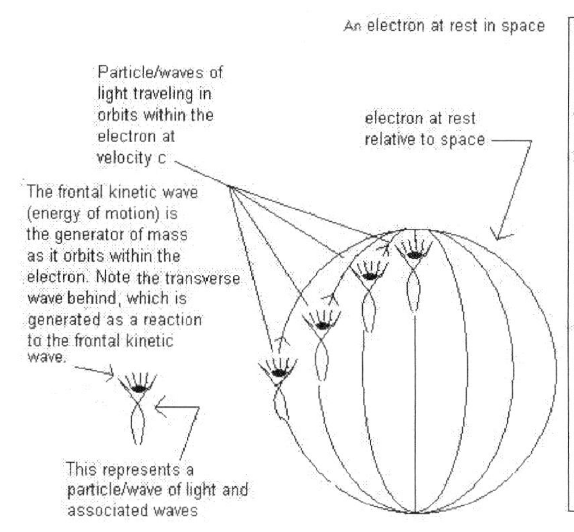

With the electron at rest relative to the space around it the Lorentz equations will not apply. The electron's shape will be spherical. As the particle/waves of light orbit within the electron they continuously propagate an angular kinetic matter wave. These matter waves are a repulsive force and space reacts to this force. This action in effect, spins the space around the electron to create what we call mass. This same action also produces a gyroscopic effect within the electron. From this configuration we now understand why an electron will travel in a swirling path as it traverses an electromagnetic field. In addition, this same action stretches the space in and around the electron to generate gravity. Gravity at this level I call "quantum gravity" and it is strong enough, at this level, to hold the particle/waves in their orbits within the electron. The quark has this same configuration.

In Figure 16. above, the reason an electron has the same configuration as a quark is because electrons are in fact quarks.

This approach answers a long standing question, "If the electron is negative in charge and we assume that its parts within must also be negatively charged, then what holds the electron together?" On page 51 in the book RELATIVITY The Special and the General Theory, there is a foot note by Dr. Einstein that reads, "The General Theory of Relativity renders it likely that the electrical masses of an electron are held together by gravitational forces." I agree with him. In fact, I call gravity at this level quantum gravity.

CONCERNING RELATIVITY AND FIELDS

I must begin by saying; "it seems that not everyone who writes about Relativity truly understands it". Hence, we sometimes get a biased or personalized interpretation as to how Dr. Einstein's relativity works in reality. After reading several different and conflicting ideas about the same subject (relativistic effects), I obtained the book Relativity-The Special and the General Theory, by Albert Dr. Einstein.

On pages 21-24, in Dr. Einstein's book referenced above, Dr. Einstein sets up a hypothetical experiment to address the idea that, relative to an observer, two distant events, in different directions from the observer, can happen simultaneously. His experiment is as follows: Take a distance along a railroad track and call one end of this distance A and the other end B. Place an observer exactly half-way between A and B, on the embankment alongside the tracks, with two mirrors (call this position M). Lightning strikes at both A and at B. The observer at position M, while holding up the mirrors, sees both flashes at the same time and concludes that the lightning strikes happen simultaneously.

Dr. Einstein reasoned, page 23, since the speed of light is constant, and the flashes of light traveled the same distance in the same time, relative to the observer at position M, then the lightning strikes, can be considered to have happened simultaneously. But note, this result is only with reference to the observer at position M and his clock.

On page 25, Section IX, in Dr. Einstein's book, he sets up the same experiment, only this time he has an observer on a train (call his position M1) which is in motion from left to right (going from A towards B) at some velocity v. Just as the observer's position M1 on the train, which is in motion at velocity v, reaches position M on the embankment, lightning strikes at A and B, the same as before. (see Figure 17.)

FIGURE 17.

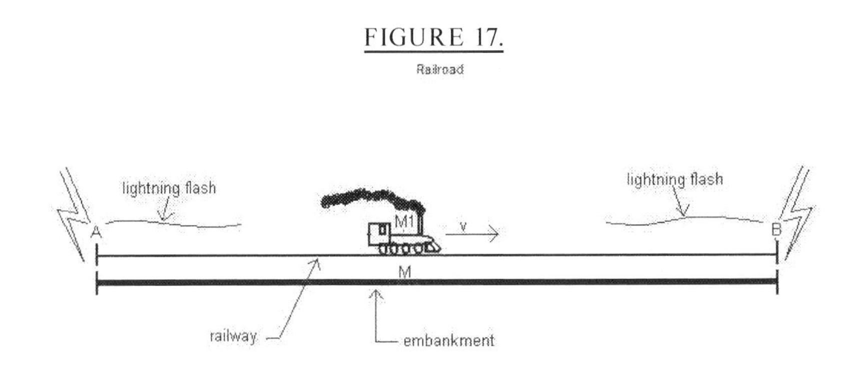

I will quote Dr. Einstein here in order to suppress any doubts as to what he meant. He says, "Just when the flashes of lightning occur, as judged from the embankment, this point M1 naturally coincides with the point M, but it moves towards the right in the diagram with the velocity v of the train. If an observer sitting in the position M1 in the train did not possess this velocity, then he would remain permanently at M, and the light rays emitted by the flashes of lightning A and B would reach him simultaneously, i.e. they would meet just where he is situated. Now in reality (considered with reference to the railway embankment) he is hastening towards the beam of light coming from B, whilst he is riding on ahead of the beam of light coming from A. Hence the observer will see the beam of light emitted from B earlier than he will see that emitted from A".

But note, this also tells us it will take a ray of light more time to catch up to you when you are in a motion ahead of the advancing ray, than it would if you were stationary. Now, we know the ray (light) is made up of photons which have wave-lengths and the tails of these waves travel at the same speed as the heads of these waves. This would then mean, with reference to what the last sentence describes, "If it takes the ray (wave lengths) more time to catch you, when you are in a motion ahead of the advancing ray (wave lengths), then it also takes the ray (wave lengths) more time to pass you." This result plays a big part in what is called the Doppler Effect. And this result is very important so we will call it (Fact 1).

The main idea, we want to retain from all the above, is the fact that, with the observer setting motionless at position M, the light ray's rate of approach is measured to be 186,000 mps or c. Now once the observer is put into motion (M1) you would think that the Light ray's rate of approach from A would become less than c (simple laws of motion). But Dr. Einstein seen it differently.

The Dutch astronomer De Sitter, by his observations of double stars, showed that the speed of light is independent of the motion of its source. Meanwhile, in the field of electrodynamics Maxwell mathematically showed that the speed with which an electromagnetic field propagates (ratio of electrical force of charges at rest vs magnetic force of these same charges in motion) was the same speed as for the propagation of light. This caused a mental leap to be made by Maxwell and the end result of this leap concluded, that light itself (energy) was an electromagnetic wave. But let it be noted, Maxwell used a physical space as the carrier (reactionary medium) for his electromagnetic field.

Dr. Einstein, under the influence of Mach and possibly the results connected with Michelson and Morley's famous aether experiment, came to two rather disturbing conclusions. Namely; space as a physical entity, does not exist (influence of Mach), and the speed of light is independent of the motion of any observer (influence of Michelson and Morley's experiments). Note that Michelson never accepted that space was void even though many seen the results of his experiment as showing just that. Note also that Newton rejected any idea where nothing (empty space) might have the ability to take up room.

In reference to the first conclusion; Once Dr. Einstein emptied space he immediately filled it back up with a physical field. Note that he replaced Maxwell's physical aether in space, with a physical field in space. On pg. 155 (Relativity-The Special and the General Theory) he says, "There is no such thing as an empty space, i.e. a space without field. Space-time does not claim existence on its own, but only as a structural quality of the field". This sounds like another aether theory doesn't it. On page 9 he says, "In the first place we entirely shun the vague word "space", of which, we cannot form the slightest conception, and we replace it by "motion relative to a practically rigid body of reference". Under this configuration, space itself does not exist but the field it is made of does. In my opinion, this is where Dr. Einstein's problems began, with reference to his attempt to unify electromagnetism and gravitation. Let me explain.

On page 285 in Dr. Einstein's book- Ideas and Opinions, he says, "Gravitation has indeed been deduced from the structure of space, but besides the gravitational field there is also the electromagnetic field. This had, to begin with, to be introduced into the theory as an entity independent of gravitation. Terms which took account of the existence of the electromagnetic field had to be added to the fundamental field equations. But the idea that there exist two structures of space independent of each other, the metric-gravitational and the electromagnetic, was intolerable to the theoretic spirit. We are prompted to the belief that both sorts of field must correspond to a unified structure of space".

It is clear from this, that he and other theorists could not imagine a single field in space being able to contain an electromagnetic field, while at the same time containing a gravitational field. The explanation for this is really very simple. Under my model space can be driven and stretched simultaneously. Also; space, within our local area in the milky way galaxy, transmits light at a certain rate (300,000 kps). Note that I do not define light and electromagnetism as being the same thing in nature. This is very important so remember this as you read further.

The reaction by space to many aligned electrons, forced into motion through a copper wire, will be to generate an electromagnetic field in the area around the copper wire. The speed with which this reaction is propagated through space has been shown to be approximately equal to the speed of light. But do not be fooled as Maxwell and others were.

Light, in a general sense, has nothing to do with generating an electromagnetic field. If we apply an electrical current to a copper wire, this will cause the outer electrons within the wire to flow along the wire in a particular direction. But this action causes a reaction within the space surrounding the wire. The electrons, once put into motion along the wire, drive the fabric of space surrounding the wire. This driven space becomes what we call an electromagnetic field. Note: you can place an opaque barrier next to a magnet which will, in effect, block any light from passing thru the barrier. But this does little to nothing in reducing the strength of the electromagnetic field as it projects itself thru and beyond this same barrier. Maxwell also theorized that the energy connected to light waves should depend only on its amplitude and not its wavelength. He, in effect, showed with this line of thought that the whole Planck unit of energy resides within the leading portion of the light wave.

But now let us address an event where we insert a gravitational field into this same space that is presently containing an electromagnetic field. Example; Let us place an electron out in space at rest and monitor its emissions. The 2nd law of thermodynamics tells us that a hot body placed within a cool surrounding will attempt to stabilize (make equal) the temperature between itself and its cool surroundings. Hence the electron will radiate energy into the space around it. This action is what we call entropy and from this we can conclude, space will eventually draw all the energy, from all objects of mass. I will talk of this later.

As you can realize here, the electron is at rest but still radiates energy, and the speed with which space transmits this radiation is the same speed as it uses to propagate electromagnetism. This is the point where Maxwell made his leap, i.e. because light and an electromagnetic wave travel at equal velocity they must in fact be the same thing in nature. But the difference between light and electromagnetism is this. Light is transmitted through space, by space, at a speed equal to c while electromagnetism is an internal spatial reaction equal to c, by the fabric making up space, to bodies (i.e. electrons) in motion through it.

Here we must recall a law connected with electromagnetic fields in that, Maxwell showed that the strength of electromagnetic fields will diminish, over distance. Radiation, on the other hand, follows no such law. Example; the earth's electromagnetic field does not reach to Jupiter but a single radio wave of light will not only reach Jupiter but will also return if reflected, while still maintaining its original speed. So how is electromagnetism and light related?

When you put electrons in motion to generate an electromagnetic field you are also automatically causing these accelerated electrons to emit light (i.e. radio waves). The electromagnetic wave force dies off in space but the light wave continues on. This says, if you could "hypothetically" stop the electrons' light emissions, resulting from their acceleration, you would still generate an electromagnetic field. And its propagation speed through the space around it would still be equal to that of light.

Faraday showed that the force configuration connected to a gravitational field, has the electromagnetic force leaving one pole of the electromagnet and then bending around to the other pole. Light, which travels in linear paths, unless reflected or under the influence of a gravitational field, cannot bend in the way Faraday describes. I believe that, a good experimentalist, by use of the laws of physics and opaque materials, would more than confirm this idea.

The question now becomes; if space, reacting to the motion of electrons, generates the electromagnetic field, then how does it also generate a gravitational field? Dr. Einstein gave us this answer above when he said, "Gravitation has indeed been deduced from the structure of space".

Follow here; the electromagnetic field is generated by the matter making up space itself, reacting to bodies (electrons) in motion through it. The speed or rate, of this reaction was found to be equal to the speed of light. Note that space's reaction rate equals c even though the cause for the electromagnetic field (electron) had a speed less than c. This would seem to produce a mystery but can be easily explained once it is understood that the super dense fabric making up space itself reacts to all forces brought against it in a similar manner. For example; as an electron is accelerated, the space around the electron doesn't react under the same rules as prescribed by the laws of motion. So even though the electron might be put into motion at a speed of ¾ c the fabric of space surrounding the electron reacts with a speed equal to c. Hence; another mystery solved.

The gravitational field is generated by the energy within these bodies of mass, stretching the space around them to create an inertial guiding path of least resistance. This gives us a picture showing that electrons impel or "drive" the space around them when they are put in motion (electromagnetism), while the energy (particle/waves of light) within the electrons, "stretch" the space around them as they travel in their orbits within the electrons (gravity).

So you can now see, Dr. Einstein was correct when he said, "Gravitation has indeed been deduced from the structure of space". But Einstein had one major problem with his perception of fields. He could not, try as he might, combine his two fields (electromagnetism and gravitational) in the same area within his "superfluous" space. Early in his career he speculated that space itself was superfluous (unnecessary). In my opinion this was his one biggest mistake and could have been avoided if he had studied Newton's work more closely. I will readdress this idea later within this work. Note above that I have energy (quanta) in orbital motion making up electrons (quark). Please keep an open mind with respect to this. I will back this postulate up later within this work.

Of further note, String theorists and Quantum theorists, are now stealing energy (virtual particles), by use of Heisenberg's uncertainty principle, from this empty space and to go even farther, Dr. Einstein himself postulated that empty space, on a universal scale, is "physically" warped by his field. How does one warp nothing?

Now let us address Dr. Einstein's second conclusion; the speed of light is independent of the motion of any observer, I have found that not only I, but others within the scientific community, have problems with this conclusion.

The Doppler effect, relative to light, is directly connected to the idea that some given quantity of light waves, passing a point (observer) in a certain amount of time (second), gives us a frequency. And this frequency can be variable depending upon the motion of the observer either toward or away from the advancing light waves.

Call this, an "empirical axiom" for it has been experimentally proven beyond doubt. But this presents a problem when we incorporate Dr. Einstein's conclusion, the speed of light is independent of the motion of any observer, into the Doppler equation.

I am not concerned with the relationship connected to the source's motion but am concerned with the relationship connected to the motion of the observer. If the speed of light is independent of the motion of the observer then so to should the distance between the waves (period T) be independent. Let me expand on this.

Given: The motion of an observer, depending upon whether he is moving toward or moving away, from the approaching light, either shortens or lengthens the distance between the individual waves of the approaching light but this configuration is only true when we factor in the observer's motion. This is how the Doppler effect is created.

Assuming the constancy of the speed of light, if we look at this light coming at us as a ray (string of waves), then the ray (string), as a whole, has to maintain the same velocity, which is c. Now, if our motion cannot effect the rate of approach, of the ray (string) coming at us, then how do we then lengthen or shorten the distance between the waves within the ray (string)? I will show the math connected to this idea later within this work. Let me say, "If the speed of light was truly independent of the motion of the observer then there would be no Doppler effect"

On the other hand; It would seem that the results of the Michelson-Morley experiment would back Dr. Einstein's idea, but because light has a constant velocity, which is independent of the source, and because Michelson left time out of his results, I'm discounting his work, and will stick with the results connected to the motional requirements of Doppler. I will address the Michelson-Morley experiment later within this work and will inject time into their experiment. As you will see, it makes for a whole different story.

Another question of importance that arises from all the above, is not that it will take more time for the ray of light to catch, and pass you, but "exactly" how much more time will it take, and by who's clock? I tried Einstein's equations and got incorrect results.

It is a foregone conclusion in the field of physics that, with reference to measurements involving speed, time and distance, you must have a frame of reference to refer to, as you do the measurements. The problem that arises from this requirement is that, according to the principles of relativity, when you compare measurements between events that use different frames of reference, and these different reference frames are all in motion at different velocities, the results of the measurements become skewed, when compared to each other.

It seems, from reading Dr. Einstein's works, he assumed that any object, not under acceleration, could be considered to be at rest with reference to the space around it. And because Dr. Einstein seen space itself as empty (superfluous), he seems to conclude that it becomes a wasted effort even to try and calculate how fast through space we are traveling. Therefore we must always choose a particular frame or body of reference and tie all measurements and results to this particular body (frame of reference).

It is now easier to understand why Dr. Einstein's concept of relativity concludes, an observer O1 on earth can say that another observer O2 on a spaceship passed him at velocity c/2 (half the speed of light), but observer O2 on the spaceship can also say that it is the earth and observer O1 that is passing his spaceship at velocity c/2. And according to Dr. Einstein, they would both be correct.

In addition, with this understanding, both observers could also conclude that they were approaching each other at speed c/4 or any other combinations, adding up to speed c/2. Mathematically, with reference to Dr. Einstein's point of view, we arrive at an enormous number of answers that can be "relatively" true for this event. But this leads to the question, "Is there one real answer?" By use of Newton's laws of motion I believe there is and I will address this particular problem later within this work, so let us move on.

The theory of relativity says, by use of Lorentz's transformation equations, when a body is put into motion, its physical length is shortened and its internal clock will run slower. (Call this Fact 2). But I must add here, "time can be a chameleon, so beware the frame of reference you choose when calculating with Fact 2." Because the idea of time and the measurement of time, is a major player within this document, I will now show another work which will give you an idea of how time and our notion of time, is viewed within this document.

MICHELSON'S DOUBLE SLIT EXPERIMENT

What do we now say about mass if light is already in motion within the electron and we assume that mass creates light (energy)? The answer to this is, "Quanta and its "frontal" kinetic wave, by moving in circles, within the elementary particles (quarks), creates the force we call mass. This says that mass is actually an angular space wave force that propagates from within the quarks.

Let me explain: A particle/wave of light (quanta) has a kinetic energy (energy of motion) connected to it by its motion through space at speed c (momentum). This energy has been empirically shown to have a frontal force connected with it (objects have been elevated by intense beams of light). As we discussed before, if we equate this energy (quanta) force to mass force, then we can say that a single particle/wave of light making one orbit within a quark creates one basic quantum unit of mass. With the particle/wave in linear motion through space we call the effect, of the frontal kinetic force, energy. With the particle/wave in orbital motion through space within a quark we call the effect, of this frontal kinetic force, mass. This configuration backs Dr. Einstein's claim, that energy and mass are equivalent (the same thing in nature).

To repeat, remember before when we found that quanta are subject to a gravitational field? This points to a conclusion that energy possesses weight and therefore, should also possess a measure of mass as Dr. Einstein postulated. But we must remember that this single quantum of mass we speak of here, when connected to a particle/wave of light being emitted by an electron, is a linear energy/mass force. It is emitted away from the electron in a straight path. What this all says is, "Quanta, moving in orbits within quarks generates mass. Quanta, moving in a straight path generate photons". As you can see, within this work and under this configuration, mass and energy in reality, really are the same thing, just as Dr. Einstein postulated.

I would hope that anyone who reads this work realizes that I didn't have the option of dreaming up a new particle or force, each time I ran into a problem. This work is based upon the idea, that the matter make-up (all particles and forces) of the universe is simply energy (quanta) in linear and non-linear motion through a medium (space).

Again, energy in linear motion creates photons. Energy in non-linear motion (traveling in orbits) creates an angular wave force we call mass. And space provides the reactive creator for all the effects (waves and forces) connected with this configuration.

Gravity, under this configuration is simply space being stretched (expanded) near objects of mass due to its reaction to of all the energy in motion within and emanating from these objects of mass. I call the gravitational field, simply an inertial guiding path of least resistance provided by the configuration (expansion/stretching) of the space around any object of mass. This configuration explains the long sought answer as to why gravity is thought to be an attractive force only. It also explains why no one has found the theoretical particle known as a graviton. Under this configuration it simply doesn't exist.

I hope it is also noticed, that several other hypothetical particles, the gluon being one, can be explained away by the results of this work. Dr. Einstein postulated that the underlying principles governing the universe, as a whole, should be very simple. I believe this work goes a long way in meeting his expectations. I realize that I repeat a lot of what I'm presenting but there is a reason behind it. A puzzle, such as I'm involved with here, takes a lot of back and forth in order to assure the right information is understood in the right order.

You, the reader might now be asking, if this model treats the universe as a solid (no voids within) then what is in the areas between the kinetic energy and the space it is in motion thru? Space is applying a huge amount of pressure upon the particles of light traveling thru it which means they are in direct contact with each other (mean free path = zero). And there is definitely grinding going on (entropy at work)

When space erupted (Big Bang), an enormous amount of the substance making up space was liberated (put into motion) and as a consequence contracted. You might say; the first and only thing all this kinetic energy, ejected from space during the Big Bang tries to do, is try and return into the space from which it came. If we assume that the output of the Big Bang was initially an ejection of plasmic energy in all shapes and sizes then we must then ask, 'how did the one big plasmic ejection break down into Planck units of quanta which could then go on to generate all the photons and mass we find in the universe around us at present? The answer necessarily becomes; ejected energy interacting with space itself led to the breaking down of all this plasmic radiation into packets of energy equal to 6.626×10^{-34} js. And since space itself is extremely dense (zero mean free paths between the particles making up the fabric of space) the liberated particles (light) are unable to be reabsorbed. Hence they are pressured (force of acceleration) to stay in motion at speed c.

This lets us say that space not only transmits light but also accelerates it. So why then, does light not accelerate to infinite speed? I will let Newton explain, for his 3rd law of motion plays here. As we discussed before, space not only transmits light but also accelerates it. This is an action so let us analyze the reaction. The pressure space applies to the liberated particle is applied to all sides or surfaces (translational equilibrium). If the particle were at rest it would stay at rest due to the equal pressures being applied to it as a whole. But because it has been put in motion (Big Bang) and because space has no room to absorb it, then space continues to squeeze (accelerate) it along. Meanwhile the space in front of the accelerated particle (light) reacts or resists this acceleration (Newton's 3rd law). The "effect" of this resistance by the space in front of the particle in motion, becomes a physical effect or what we call "energy".

Take note here that what I've just described, is happening within the expanded area of space (gravitational field) around where the Big Bang occurred. If a liberated particle (light) does not become a participant in the making of a mass object (quark, nucleus, electron, or neutrino) and then finds its way out beyond all the gravitational fields where space is normal (much more dense), it will then be eventually shredded from the photon it is part of and then become reabsorbed into space itself. Hence it comes to rest, expands, and becomes part of space again and adds more pressure to the fabric of space itself. As we discussed before, after enough orbital energy (mass) is converted back to linear energy (light) and reabsorbed by space, then the particles of space under all this added pressure goes into annihilation (Big Bang) again somewhere else.

Because, relative to this work, all elementary particles (quarks) consist of particle/waves of light in motion within them, there ought to be a repeating and periodic mass wave being propagated from these particles. Let us analyze this.

Let us first set a diameter for an object of mass (we'll use an electron). Using the Fermi length as a base size, we will give (guess) the diameter for an electron to be approx. 10^{-13} cm. Using π we find the circumference of the electron to be 3.1416^{-13} cm. This circumference will be our distance d. Our rate r would of course be c since we are calculating the time it would take for a particle/wave of light to complete its orbit within the electron.

Now we plug this data into the well known equation $d = rt$. The result is 3.333^{-22} seconds. This says that reactionary mass waves will be propagated into the space surrounding the electron every 3.333^{-22} seconds, relative to an observer placed outside the electron. Hopefully at some point in the future an experimentalist will confirm, not the sizes I use, but the idea itself.

I am in no way saying, that the experimental findings will exactly match my guess as to the true diameter of an electron, but if the time between the propagated mass waves can be determined this would automatically give us the diameter for the electron, (relative to earth time). Note that within this model no body of mass is seen to have a radius of zero.

We must also realize, as the electron itself, is put into motion, and because light has a constant speed, the orbit times for the particle/waves of light within the electron, will increase. This is because the particle/waves which move at only one speed c through space, must share this speed between the orbits they're in within the electron and the motion of the electron itself thru space.

Hence, it would take the particle\waves of light far longer to complete an orbit within an electron if that electron has a velocity approaching the speed of light. This will of course, effect (lessen) the electron's mass and inertia and also its length. But, in reaction to this loss of mass and inertia, a kinetic wave will form ahead of the electron (energy of motion) or deBroglie's pilot wave if you will.

In addition, because the orbit periods become longer, the internal action of the electron becomes less. And not just the electrons will experience this effect but also all the other quarks within the atom will to, which means the atom as a whole will be less active internally. Hence molecules will be less active between each other. Hence molecular action will be slowed (Lorentz). What does Dr. Einstein say about this? Fact 2. The theory of relativity says that, when an object is put into motion, its physical length is shortened and its internal clock will run slower. Lorentz theorized this effect while I "mechanically" explain it.

Some quantum theorists believe that the electron has a radius of zero, but I don't believe they realize the seriousness of the impact this would have on their mathematics, when trying to do calculations at the quantum level. If you do away with π, you become unable to factor the internal physical characteristics for any particle of this type, you are working with. I also have a problem with the idea, a physical object can exist without taking up room or area (point). This is the main reason I introduced the fifth dimension (size).

The fifth dimension has five levels. The first level is the quantum level (10^{-24} cm). The second level is the classical level (1 meter). The third level is the astronomical level (AU 149,600 km). The fourth level is the galactic level (1 light year) and the fifth level is the universal level (infinity).

As an aside: In physics, work is defined as the measure of energy transfer that occurs when an object is moved over some distance by an external force. This says that energy and force are directly related. This points to a conclusion that says, "all forces can be factored in Planck units of energy". We must start thinking in plank size.

As I said before, within the spherical electron each particle/wave of light is in motion in a circular pattern which causes the <u>frontal</u> kinetic energy force, tied to the particle/wave, to be propagated outward from the electron by space, at an angle. This is mass (also known as angular force waves or matter waves or quantum force).

By now I am sure that some are questioning my use of the term frontal kinetic energy force. This is a kinetic wave (DeBroglie's pilot wave) that extends for some distance ahead of the particle/wave of light or the electron when in motion. This configuration explains how the particle/wave knows whether one slit or two is open relating to Young's double slit experiment. Quanta (light) has been proven, to contain both a particle and a wave yet some refuse to accept this fact. Simply put, "A particle of light, while being transmitted through space at speed c, generates a frontal kinetic space wave which in turn, results in a transverse space wave behind." The particle in motion through space is the action. The frontal kinetic wave generated by the particle in motion through space is the reaction (spatial effect). And as I showed before, this frontal kinetic wave generates a spatial quantum force. We call it energy.

In reference to Young's double slit experiment, if one slit is open the frontal kinetic wave hits the slit first but does little to cause the particle/wave that is following, to deviate from its path, hence the particle/wave goes straight through the single slit and collides with the photographic plate behind.

But when two slits are open the frontal kinetic wave arrives at the slits ahead of the particle and spreads out behind the slits into what I call sister waves (interference), which then affects the path of the particle coming behind. Hence, the particle is deviated from its linear path after it passes through the slit. (See Figure 18.) I can't remember which book I read this in, but Dr. Einstein postulated that maybe a ghost wave is responsible for Young's Double Slit phenomenon. But we must remember that Dr. Einstein seen space as superfluous (nothing) so he was restricted by his own theory from possibly finding an answer. In my opinion, he was looking right at the prize but didn't recognize it.

FIGURE 18.

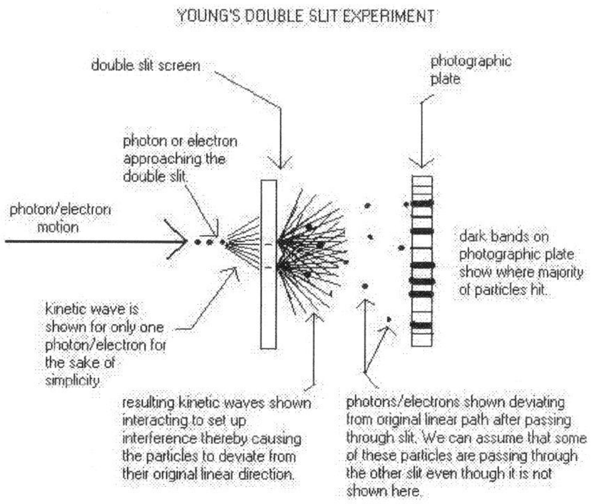

YOUNG'S DOUBLE SLIT EXPERIMENT

But why did Dr. Einstein see space as empty? The answer, believed by most, is the famous Michelson/Morley experiment, performed in order to detect the earth's drift (motion) through what was then known as the aether (space). So let us analyze this experiment.

Michelson designed a device called an interferometer. This device basically had two reflecting mirrors placed at right angles and at "presumably" equal distances from the beam splitter. I say presumably because Lorentz's equations relating to foreshortening, would have come into play here. But this will be of no concern to us in respect to what we will be looking for, which is times.

The light beam was split and the beams went off at right angles to each other and then converged once again. Michelson expected a phase or velocity differential between these beams, due to the earth's motion through the aether (space), when they were rejoined. He was denied the result he sought. The beams rejoined in-phase.

But what about the time it took for each beam to rejoin. Therein lays the answer. Michelson couldn't calculate, due to the small distance between the parts of his setup, the time it took for the individual beams to rejoin.

So we are left with an experiment involving motions but no times for these motions. This to me, is an unacceptable experiment. (review work on page 78)

THE QUANTUM OBSERVER

At present the scientific community sees gravity as a very weak force at the quantum level that still somehow manages to hold galaxies and groups of galaxies together. Gravitation is thought to be of no importance in the interactions of atoms and sub-atomic particles. The "assumed" force of gravitation reaches great distances, while nuclear forces operate only over extremely short distances and decrease in strength very rapidly as distance increases. The supposed exception to this line of thinking is a theorized particle known as the gluon. Theoretical physicists, as a last resort, had to sprinkle some magic (gluons) into their equations to preserve the sanctity of their standard model. This boson was assumed to be real in nature until the standard model was revamped at some point after my demise. This was when true quantum reality set in.

It is theorized that electromagnetic forces between particles can be repulsive or attractive depending on whether the particles both have a positive or negative electrical charge. These attractive and repulsive forces are seen as canceling each other out, leaving only a weak net force. It is thought that because gravitation has no repulsive properties, no such cancellation or weakening occurs. Out of the above information note that gravity has no repulsive or neutral characteristics. The question must now be asked, why all the wrong assumptions in the past? The answer is scale. Scientists at present try to describe reality at the quantum level but they are still doing so using classical thinking. So we will now correct that error.

There exists one particle, whose interaction with itself, in two different states, is responsible for the physical makeup of the universe and the movement it contains. This particle can exist in two states depending upon which transformation it has undergone. Without momentum it expands, comes to rest and actually becomes part of the fabric of space itself. After the next transformation (Big Bang) it gains momentum and contracts to become a kinetic particle of light and then goes on to create nature around us.

The kinetic energy I have just described is the direct generator for the "assumed" force we call gravity. As particles of light, very small, are transmitted under pressure through space, they actually, by way of radiation pressure, expand physical space. This is due to space reacting to the individual particles of light in motion thru it. The particles of light, which are greatly contracted and very dense, physically drives and expands the space ahead of them. This physically driven space is in the form of a kinetic wave. Again; in addition to space being physically driven it is also physically expanded within this driven space. This is the mechanical cause for the effect known as gravity. It is not so much a field as it is a "spatial domain" containing an inertial guiding path of least resistance for any object that enters it. This says that gravity, when defined as an inertial guiding path of least resistance, will affect mass matter and energy in a manner different, from how forces do. Using this same scenario but applying it to something like the sun, we find that the space within and around the sun becomes expanded to a degree that is almost unimaginable. Imagine the trillions and trillions of light particles emanating from something like the sun and allow these light particles to generate a tiny gravitational field (expanded space). Multiply trillions and trillions by tiny and you now have a considerable amount of expanded space within and around the sun. And a man named Newton figured out how to mathematically measure the effect of this expanded space (gravitational field) by using what he called the Inverse-Squared Law. Also, Einstein's space-time continuum, which he proposed as a field of geodesics "shortest possible paths", could represent my expanded space though somewhat inexactly. Einstein curves his space-time, though he could never physically explain how, while I simply expand space using kinetic energy. Note that Einstein could never unify his curved space-time with the forces while my mechanical explanation for gravity shows no need for such a unification. This is because, under my explanation, gravity is not an active force at all but is in fact an inertial guiding path of least resistance.

Imagine a space ship with its engines turned off, traveling through space far out in the solar system where space would be much more compact and normal and then approaching the sun where space becomes more expanded the closer you get to it. There is a law of motion, which says that all moving objects, unless subjected to an opposing force, will tend to take the path of least resistance. The space ship, being a moving object, discovers a path with less resistance and it leads toward the sun. This is a consequence of entering the gravitational field belonging to the sun. Does it now become obvious as to why no one has been able to confirm the existence of the theorized particle called the graviton? It simply doesn't exist.

The differences in Newton's, Einstein's and my view of the gravitational field are as follows:

Newton held that a massive body such as the sun, permeates or fills the space around it with a gravitational force that makes the planets move along curved trajectories instead of straight lines. This approach to explaining the gravitational field held for centuries without challenge. If Newton had known that later experimental data, showed his theory to be "not quite right" in predicting the seconds of arc that mercury's perihelion changes every year, I'm sure he could have solved for the difference, and just as convincingly as Einstein did. That's the genius of being a genius isn't it?

Einstein postulated that the sun's "mass" warped space-time so that the planets move in a curved trajectory also, but not because the sun attracts them, but rather, because they are moving along geodesic routes, or the "shortest possible paths" in space-time. A mathematician named Grossman showed Einstein how to apply Riemann tensors to his theory in order to relieve him of the impossible task of trying to describe his gravitational fields using the two dimensional, mathematical tool called Euclidean Geometry. Einstein basically created a three-dimensional grid replica of space without a gravitational field and then mathematically applied the effects; a gravitational field generated from a massive body like the sun would have, to this grid. He called this tool he built a space-time continuum and stated that what was being warped was not actually space but this continuum, which was basically a configuration of a gravitational field. So now he was relieved of the problem of trying to convince everyone that what was actually being warped was space, which by now was considered to be void of any physical matter, which excluded warping of any type being done to it. We must understand that at this point Einstein had been convinced by a flawed experiment (Michelson's), as many others had also and still are, that there existed something made of nothing. (See Axiom # 2) And this something was the space within and all around us. You must also understand that Einstein, like myself, did not see the cause for gravity as an active force. The scientific community believed that the gravitational field would eventually be explained by a "force carrying" particle or wave property. Newton and Einstein, even after presenting their individual theories on the gravitational field, still could not explain the exact physical mechanism that causes it.

I will now present my views on both, the gravitational field and its exact cause. I shall begin by admitting that I used parts of Newton's theory and also parts of Einstein's theory in order to complete mine. From Newton, I used his view of a physical space in which to place my gravitational field, though my space is not an absolute non-moving entity such as he believed his was. I have undulations of pressure constantly in motion throughout space. I also used his view that the sun's energy affecting the space around it creates the gravitational field. The difference being here is that I have all the particles of light being emitted by the sun, stretching or expanding the space around it, rather than permeating or filling it. From Einstein's theory I borrowed his space-time continuum and my space continuum shows a stretched/expanded configuration of space around the sun rather than a particle generated attraction field. My Gravitational Domain (space continuum), which actually includes physical space, works exactly like Einstein's with the exception being that mine goes on to explain the actual cause of the effect known as gravity. I allowed a physical space, within and around the sun, to exist within my setup and then I showed its graphical representation using Riemann tensors. This shows very clearly that the effect of gravity, generated by the stretching or expansion of physical space around massive objects, will provide an inertial guiding path of least resistance for any object traveling near it, including light itself. A physical representation of my Gravitational Domain would be analogous to taking "two" rubber sheets, stretching them tightly to each other, and then putting a baseball between them. The sheets would be expanded the most; the nearer you approach the baseball. This is where the "inertial guiding path of least resistance" would be the greatest. In measuring the gravitational field surrounding the earth you will find that the inverse square law works only until you reach the source (earth). Inside the earth the equation becomes log-rhythmic all the way to the center. Using the Laws of Physics to back my views on this argument I can assert that an inertial guiding path of least resistance can alter a shortest possible path. This means that the path of least resistance can cause the distance of the shortest possible path to increase while the inverse is not possible. Only a force can cause a body in motion to deviate from the path of least resistance but both, a path of least resistance and a force, can alter the shortest possible path.

THE QUANTUM FORCE

The quantum force within this model is the repulsive action between "angular" quark matter waves (mass). This force keeps the particles of light within the quarks from coming into actual physical contact, which would lead to mutual annihilation. The quantum force is stronger than gravity at this level in nature, else gravity would cause the particles to come together and the universe we live in would be nothing more than the remnants of continuous annihilation. Because particles of light in a "piled up" configuration, became scarcer, as the quark building period continued; various types of quarks were created. I use only three quarks in showing how the universe is configured; the three heavier quarks that make up the proton, the medium quarks that makes up the electrons and the weak/neutral quark that makes up the neutrino. I do not suggest that the quarks above are the only quarks but only show how a quark configuration would work in describing the makeup of the universe.

Here is a fine point that I need to show. The quantum force (angular matter waves) overpowers quantum gravity and keep the quarks from annihilating, but the strength of these matter waves dies off quickly with distance and gravity once again takes over. This keeps constant pressure upon the matter waves. And the presence of this continual pressure has an effect we know as entropy. Working, working, working, to rob all mass of the kinetic energy that generates it. This mechanical action is also the cause for the phenomenon involving the "perceived" loss of mass, when nuclei are formed by constituent nucleons. Matter waves (angular force), when being applied to each other will interfere and this wave interference equates to a loss of angular force (mass). This indicates that mass has to be a wave property.

First came the densest of the quarks, which were created when particles of light were in abundance, immediately after the Big Bang occurred. These were the quarks with all possible orbits containing the maximum particles of light allowed. These particles also had the strongest angular matter waves.

Next came the creation of the intermediate, weak, and weak/neutral quarks, as particles of light were becoming scarce. The angular matter waves associated with these particles were of various strengths, depending on the number of light particles in their makeup, but were still repulsive to one another.

Basically, quarks are gyroscopic in nature. The particles of light traveling in their orbits within quarks, while exerting their gravitational pulls upon each other, produce what Aristotle called impetus and what we have come to know as gyre/inertia. This is the mechanical cause that gives matter the tendency to remain stopped when in a stopped position, and also to remain in motion when in motion.

In addition, depending upon the orbital direction that particles of light are moving within the quarks, one will be the anti-particle to the other. Within this model, if any quark is vertically rotated approx. 180^0 it becomes its anti-particle. Though it has yet to be proposed, there does truly exist an up and down direction for all the quarks generated by the kinetic energy that was ejected by the Big Bang. There is one exception, the weak/neutral quark, which I shall introduce shortly.

This might seem to make for a very fragile configuration, but we must remember that the gyroscope is a very stable instrument that requires a significant amount of "directional" or "vector" force (high-energy) in order to change its orientation in space. Under normal conditions the heavier quarks, within the particles they are forming, do not "vertically" reposition themselves in relation to each other. Chaos would be the order of the day if they did.

During the mass particle creation period, the quarks that were created with an excessive amount of particles of light on one side of their mutual axis did not exist for long, as this configuration would necessarily cause the quark's axis to rotate, which would lead to destabilization. Force symmetry must be in balance throughout the quarks else an ejection of kinetic energy or annihilation is in the offing. Because of the chaotic process that quarks were created by, I would hazard a guess that a good portion of quarks experienced annihilation before they could become the more stable parts of atoms. I would further venture that much of the residual radiation that was predicted by Gamow, Alpher and Herman and was later discovered by Penzias and Wilson, is actually the remnants of the huge amount of annihilation that occurred just after the Big Bang.

The fact that background radiation is mainly in the microwave, infrared and radio wave lengths speaks of annihilation. When photons are ejected from normal quark decay they group together in a very orderly fashion and move off in a linear direction. The photons created from annihilation would necessarily show the results of a chaotic birth with very little order. Hence, most of the radiation would not group up in large numbers, as they would be undergoing a violent scattering.

Now let us investigate the quark that allows the neutron to be considered neutral. This quark was formed towards the end of the quark creation period. This means that this particle would have close to a minimum number of orbiting light particles in its makeup. The uniqueness of this particular particle is in its being both left handed and right handed, simultaneously. This particle might be considered as an extension of the Pauli Exclusion Principle for it contains approx. equal numbers of light particles that orbit in opposite directions. In addition, due to the lesser amount of particles of light within, it would also generate a very small amount of matter waves (mass). As long as the kinetic waves, from these opposite traveling particles of light do not interfere with each other they can exist within the same quark. This causes two different matter waves to be generated from one quark. This allows this particular quark to be compatible with all other quarks, as it would have no anti-particle. When this neutral quark is approached by another quark, no matter the configuration of this approaching quark's matter wave; it is both attracted and repulsed at the same time. In addition, this neutral quark, because of its makeup, basically exists in two configurations. The first configuration, combines it with an electron to become what I call a "nulectron", and this combination of electron with a neutral quark in orbit around it, allows the electron, which is now neutral, to take up an orbit around a proton in the nucleus. This gives the proton what is seen as a neutral charge and it now becomes a neutron. The second configuration relates to this neutral quark after it is freed from the atom it resided within. Because of its sparse makeup it has a very high "penetrating" capability. In the realm of high-energy physics, this neutral quark goes by the name, neutrino.

Another capability the electron-neutrino particle (nulectron) possesses, is that it can possibly be exchanged between protons. The nulectron is an out-rider to protons hence we can say it takes two separate particles to create a neutron (proton/nulectron) within the nucleus of atoms.

With this neutral quark mediating, we could possibly have a proton and its anti-proton in close proximity within a nucleus without them annihilating each other. I find it more than a coincidence that in almost all stable nuclei, the number of protons are approx. equal or less than, the number of neutrons. Following the rule, that most things in nature exist in their simplest form, I would say that the neutral quark (neutrino) is the only reason annihilation did not happen to all the nuclei created after the Big Bang.

As an example, let us examine the alpha particle. The alpha particle has two protons and two neutrons in its makeup. The two protons naturally repulse each other but with two neutrons between them they are allowed to remain in close proximity, and can then build larger and more complex nuclei. The neutrons matter waves (quantum force) keep both neutrons and protons they come into contact with at bay so that annihilation can be avoided. Quantum gravity also plays here to hold the nuclei together. This configuration is so simple it has to be possible.

Particles of light, traveling in orbits within quarks, angularly drive space, as they move around their orbits. This driving of space by the light particles sets up an angular wave in space (quantum force matter wave). This means that the quark radiates matter waves in 360^0. The matter waves sweep out into the space around the quark in an expanding configuration. These waves that are angularly radiating out from the quark meet up with other waves just like them and, having driven space meeting each other head on, they repel one another. On the other hand, if the second quark has angular matter waves opposite to the first quark's matter waves, quantum gravity can then cause them to join together and annihilate each other. Galactic interactions seem to show these same traits though on a much larger scale.

The reason atoms do not fly apart is because the nucleus has a huge number of light particles within it and, as I pointed out previously, the light particles making up the quarks are what generate the gravitational fields responsible for what we call gravity. Quantum gravity keeps the quarks from flying out of the nucleus while the quantum force keeps them from coming together and annihilating each other.

I realize this description I am giving is not quite, in accordance, with accepted theory but there is reason behind it. Within this model space is physical and actually made up of particles of light at rest (dormant energy). In this model, space can be put into motion in four ways. Excluding the Big Bang; you can drive space in 360% such as the kinetic energy traveling in orbits within quarks do which makes space produce an orbital angular wave such as I have just explained with matter waves (mass/quantum force). You can drive space as I explain in detail under the section labeled electromagnetism to create the electromagnetic force. And you can, using kinetic energy either inside or outside the quarks, stretch or expand space which generates gravitational fields. Note: The nuclei and the atoms have two physical characteristics within them, matter waves (the quantum force/mass), and gravity, which interact to keep them stable.

PART FOUR – GRAVITY

CONCERNING GRAVITY

Concerning gravity we must go back in time to the period right after the Big Bang when gravity, with reference to our local area of space within the universe, first came on stage. According to the Big Bang theory the only action that would have been taking place at this point in time was a huge outpouring of plasmic radiation. My model agrees with this hypothesis for this is the point in time when gravity materializes. Plasmic radiation within the huge outpouring of energy from the rupturing of space, eventually began to separate into photons. Within this work, radiation affects space as it travels thru by expanding its fabric, which generates what I call, "an inertial guiding path of least resistance" (gravitational field). Hence, the first force was perpetual space pressure, the second force was radiation pressure and the third "assumed force" was gravity as an effect of kinetic radiation. I say "assumed force" for I believe that gravity is not a force at all but a physical configuration of space itself. Einstein said more or less the same thing except he discarded any need for a physical space but then went on to apply fields to what he called space-time. In effect, he did away with what was called the aether for an all-pervading entity called fields. An analogy to this line of reasoning would be to deny the existence of brown horses while theorizing the existence of white horses.

To continue; as the particles of light ejected from the Big Bang pile upon each other this causes a low pressure (high gravity) condition that allows particles of light to physically force themselves into orbits around each other which in effect, generates quarks. In the creation of quarks also comes the generation of gravitational fields around the quarks. This configuration explains why quarks have no nucleus.

As an aside; can you see that what I've just presented solves for the magical action known as "the instantaneous acceleration of light"? The light was already in motion before it was emitted from the quark, proton, neutron, electron, etc.... Let me repeat; the light was already in motion before it was emitted. I'm now ready to discuss what gravity, which has been hiding behind the curtain, really is. I will begin by stating, "Gravity, in reality, cannot be acknowledged as an active force".

Gravitational theories have been produced by such as Newton and Einstein that mathematically describe the effects of gravity. Quantum theorists are also in the ring by way of a theory that includes the graviton. Some time ago I ran across a theory that posited that the reason a cannonball and a feather, within a vacuum, will fall at the same speed is due to the cannonball's greater amount of mass-inertia helping the cannonball resist any change in its motion. Hence, the cannonball responds to the force of gravity at a slower rate than does the feather which contains less mass-inertia and this causes them both to fall at the same rate. I tried to apply math to this idea and it failed almost immediately. The problem with this explanation is as follows; the gravitational field would necessarily have to know the difference in masses between the cannonball and the feather in order to work the way that is stated above. I would think that this mass-inertia based theory would plow up several problems when applied to photons which are known to have no mass/inertia. Note: the effects connected with gravity have been described mathematically but the actual physical-mechanical cause for gravity is still in hiding. I intend to change that within this work.

In my opinion, Einstein came closest with his geodesic spacetime model but he ended up tied in mathematical knots with tensors and shortest possible paths while also trying to include a mysterious particle within his work that goes by the name of graviton. In all that I have read I've never known him to reference anything like an inertial guiding path of least resistance for gravity.

I will state, the effects of an "applied" force relating to acceleration, is not the same as the effects of an "inertial guiding path of least resistance" relating to a gravitational field. Acceleration is the result of an applied force while a gravitational field is the result of kinetic particles of energy (light) "asymmetrically" expanding the fabric of space (domain) they are traveling thru, which can then provide an inertial guiding path of least resistance for any bodies of mass or energy that might happen by. Note that I have yet to and never will, describe gravity as an "active or applied "force.

Acceleration is the result of a force being applied (bat hitting ball), while gravity (an inertial guiding path of least resistance thru space itself) can be defined as physical space being physically more expanded the closer one gets to the source of the gravitational field. You might say; as nonaccelerated objects travel thru space they can't help but take the easiest routes. Any other approach requires adding energy or propulsion if you will. So let me now state; with reference to "forced" acceleration, a body must physically interact with an activity outside of itself. With reference to gravitational acceleration, this same body has no choice but to allow its internal forces (kinetic energy) to be brought to bear against the gravitational field (directional/ asymmetrical configuration of expanded space) it is residing within.

I cannot understand why Dr. Einstein, surely knowing there had to be "something" different between acceleration (applied force) and gravity (inertial guiding path of least resistance), ended up calling them equivalent. The mathematical results might seem to be equivalent but the physical mechanical factors involved in producing these equivalent results are very-very different. It is well known that an accelerative force will cause electrons to emit radiation but gravity produces no such effect upon these same electrons. This fact allows me to say, "the effect of gravity upon objects is presently seen as an acceleration produced by a particle called the graviton, but in reality the object, once entangled within the gravitational field accelerates itself. As an example, if you drop your keys which then end up on the floor you can truthfully say, "the keys accelerated themselves towards the floor".

Quoting Einstein, "There is a remarkable property belonging to gravity such that, bodies that are put in motion, through only the influence of a gravitational field, receive an acceleration that has nothing to do with either the weight or mass of the object in motion." It becomes obvious that he "eventually" might have thought the gravitational field was applying "force" here to accelerate objects that come into contact with it. But under my model the gravitational field does not act upon any form of energy or matter. In fact, energy can both generate and act upon gravitational fields. Only by use of a physical space is this possible.

I would think that I'm supposed to inform you at this point that the inverse square law of attraction applies here but that would be a mistake. The problem is not with the equation but with the term "attraction" that is tied to it. The term "attractive force" speaks of an applied accelerative pulling force. Gravity under this model is generated, not by a mysterious attracting particle (the graviton) but by a directional/asymmetrical expanded configuration of space itself. And radiation is what creates this configuration within space. This effect of space being expanded (gravitational field) is a direct result of "physical" space reacting to kinetic energy (radiation) in motion thru it.

Note: under the requirements of thermodynamics we know that all matter radiates. All of matter's internal motions, namely the kinetic particles of energy (light) traveling in their individual orbits within the various quarks making up matter, simply find it easier to work their way down thru the more expanded space toward the source thru an inert, directionally expanded space, than it does thru the more normal denser space above. Hence, we are not "pulled" down by any earth generated attractive force but rather, the kinetic energy within us, by its own actions within the gravitational field, forces itself to take the inertial "guiding" path of least resistance. It's really just that simple.

Keep in mind that within this work, the physical configuration of all gravitational fields has space being expanded to a greater degree the closer we get to the source of the field. This is why I use the terms asymmetrical and directional when describing the physical configuration of expanded space surrounding bodies of mass. This spatial configuration provides an inertial guiding path of least resistance to the individual particles of light traveling thru their individual orbits within the quarks, which then guides the quarks and the body of mass the quarks are within, toward the generator (body) responsible for the gravitational field they have encountered.

To give an analogy, imagine an out of control vehicle traveling down the freeway. The out of control vehicle would, under normal circumstances, end up somewhere off the freeway but, we find in this case and due to the guard rails alongside the freeway, the out of control vehicle is subjected to an inertial guiding path of least resistance (the guardrails) which allows the vehicle to take the inertial guiding path of least resistance, namely the freeway. Note that it was the vehicle (kinetic energy) that acted upon the guardrails and not the guardrails (gravitational field) that acted upon the vehicle.

Again; gravity is not an "accelerative" force in any real sense of the word but is in fact, what we might call, an "inertial guiding" or directing path of least resistance. The inertial asymmetrical gravitational field is just sitting there in space and is being constantly replenished by the radiation from its source. We run into it and by way of our internal actions, we force ourselves down the inertial guiding path of least resistance it offers to us. If this work does nothing more than explain the physical-mechanical difference between the effect of gravity and acceleration I will be more than satisfied. Let us now engage with what is called weight.

The reason a ping pong ball and a steel ball bearing will fall at the same rate is because asymmetrically expanded space (the gravitational field), cannot offer a "guiding path of least resistance" to an object as a whole, but to the only real thing it sees, the individual particles of light which make up the quarks which make up that object. With that said let me assert: Only the particles of kinetic energy that makes up all matter can be considered to possess a characteristic we call weight. As I showed before; it must be understood that the gravitational field contains no active force components but can be best described as an inertial guiding path of least resistance within space itself. In other words, the gravitational field does not physically pull (attract) anything, it can only present to objects it comes into contact with, an inertial guiding path of least resistance that ends up at the energy source (body of mass) that generated it.

We (objects of mass) feel the guiding result of gravity when we come into contact with a gravitational field because the kinetic particles of light in orbital motion within the quarks making us up, find it easier to try and orbit in a direction that will eventually take the particles to the source of the gravitational field (earth). Keys if dropped will react in the same fashion. The particles of light will find it harder to try and work away from the source of the gravitational field they have come into contact with, hence the force they apply to the field ends up forcing themselves to follow the inertial guiding path of least resistance which leads to the source of the field. What I've just described doesn't apply to just us humans but to all of creation itself.

In putting all this another way, gravitational fields are generated by the kinetic energy that radiates out from all objects of mass. All matter radiates hot to cold, hence the earth radiates out into the colder space surrounding it. This action causes the fabric making up space itself to expand and the fabric of space closest to the earth expands the most (less dense/less pressure). This means, of course, the fabric of space farther away from earth is less expanded (more dense/more pressure). The end product of this is what we call a gravitational field.

Let me reiterate; the gravitational field does not act upon energy or matter. In fact, it's the internal actions by the kinetic energy within matter that acts upon the gravitational field. Quoting from the Encarta encyclopedia; "Galileo had discovered that heavy objects and light objects fall with the same speeds. Aristotle was sure that heavy objects would fall faster, although he never tried it. We now can see why Galileo's discovery is true. The Earth pulls harder on a heavy thing, but the inertial mass of a heavy thing is higher; the two effects cancel out, and all bodies fall with the same speed". I disagree with this explanation and I dispute the idea that the earth or its gravitational field, are pulling on anything.

Note that at present, gravity is supposedly generated by a mysterious and invisible particle called a graviton. This graviton has a special property that allows it to come into contact with objects of mass and cause them to move in a direction opposite from the direction the graviton is moving. We must assume, under this guise that the graviton, which evidently has no momentum, meets an object and then it applies a reverse force to cause the object to be put in motion towards its rear. If you question my physical gravitational field, then compare its possible existence, to the possible existence of this graviton thingy. Not only that; try to introduce the idea of a black hole absorbing all the kinetic energy and the mass it generates, from the objects it entraps while the gravitons are left untouched to then gather together and increase the gravitational ability of the black hole itself. I will now give you a gentle warning; don't bet the farm on black holes. The reason I say that is because gravitational lensing seems to be absent with respect to objects in motion behind and around the supposed black hole at the center of our galaxy. If black holes truly exist and do possess the enormous force of gravity as theorized then gravitational lensing should be showing its effect everywhere around the black hole. Just saying.

With this in mind, we still must address the problem of values of weight changing the further the object being weighed gets from the source (earth) generating the gravitational field. If we assume that the particle called a graviton exists then we must assume that the gravitons radiating out from the earth eventually become more spread out the farther they radiate (pi). This seems to tell me that it might become possible for a very small object, say a pin, to avoid gravity altogether if it can stay away from the graviton's path thru space. Under this scenario one pin could have some value of weight while a pin right beside it would have zero weight. This particle called a graviton has yet to be found. Need I say more? I could, stop right here for the paragraphs above contain all that needs be said concerning this subject but because some might need more convincing I will continue.

Note, even though a ton of lead has many more particles of light making it up than a feather does, the gravitational field (asymmetrically expanded space) offers its "inertial guiding path of least resistance" to both of these objects, individual light particle by individual light particle, so that they will fall at the same rate. "We assume a vacuum".

Now some might question; is it the particle property or the wave property that gives light its weight? It's the particle property of course, for the wave property connected to light is nothing more than "space" reacting to the particle in motion thru it.

Note: a particle of light in motion thru the fabric of space generates a frontal kinetic wave and in response, a trailing transverse wave. Hence, it's the particle portion of light that possesses the property we call weight. In other words; the gravitational field (a configuration of expanded space) cannot offer an inertial guiding path of least resistance to itself (space) but only to the individual particles of light making up the photons and quarks which make up all mass-matter. To repeat what's been said before; the universe as a whole contains only dormant energy (space) and kinetic energy (light).

A single particle of light has momentum and weight but no gyre/inertia or mass. But we must keep in mind that out in space beyond the gravitational fields, weight would be an unknown property but mass would still exist. What this says is; any object, no matter its size or makeup, will weigh zero when no gravitational field is present. The reason a photon has no mass is because it consists of single particles of light, which bind together and are all moving in a straight line. It takes particles of light traveling in orbits within quarks to create the gyroscopic effect that gives an object its gyre/inertia and also its externally located property of mass. This means that the gyroscopic value (angular force) for one particle of light, moving around its orbit, within a quark, at speed c, is the basic unit that must be used when calculating the true absolute-mass for any particle or object. At present quantum mechanics is unable to confirm this hypothesis but what I've presented within this work goes a long way in making that possible.

In adding to what is being shown here; a photon has no quarks within it, and as such, has no gyroscopic action, which means it has no inertia/mass or spin. But particles of light can revolve and this action can cause a change in polarization. I remember reading a science book years ago which stated that most of the light coming from the sun and entering earth's atmosphere becomes mainly polarized to just one plane. But now let us return to gravity.

A body such as the earth should tend to ride on the outside of its orbit within the Sun's gravitational field "inertial guiding path of least resistance". This is due to the earth's properties of mass/inertia fighting the sun's gravitational field as it falls around it. I read a physics book once that proposed that flight time would be equal for an arrow dropped from say 4ft and an arrow shot parallel to the earth at 4ft. Given that gravity doesn't recognize an objects mass in either case here, then we have in one instance, an added physical parallel value of inertia/gyre momentum. This makes me question the book's hypothesis and also gives a possible explanation of why Newton's equations had trouble with Mercury's perihelion. Just a thought.

Quoting Einstein again, "There is a remarkable property belonging to gravity such that, bodies that are put in motion, through only the influence of a gravitational field, receive an acceleration that has nothing to do with either the weight or mass of the object in motion." Even though I disagree with his main premise this still tells us that the gravitational field doesn't recognize the total weight or mass of an object. This leads us to the question, what part of matter does the gravitational field actually interact with? The answer; only what is different from the field (space) itself, which can only be, the individual Planck particles of light.

We know from above, that the gravitational field doesn't see the total mass or weight of galaxies, stars, planets, cannon balls, feathers, atoms, electrons, or quarks, but something more fundamental within these objects. So now we ask ourselves, what is more fundamental then neutrinos, electrons and quarks? The answer partially shows itself to us as the component of high-energy physics known as annihilation. When a particle and its anti-particle are brought together, all physical properties within the particles are ejected as energy (gamma rays). Since we have found that gamma rays are simply photons of very high frequencies, we now must ask if photons as a whole are what the gravitational field will interact with. The answer is in the negative. It has been shown that photons, no matter their frequencies, will all bend to the same degree in a gravitational field.

So what is left to analyze? The answer is the single quantum of energy itself. This single quantum, which is the smallest amount of substance in the universe, is what the gravitational fields "interact with" but "do not act upon". Again; note that the gravitational field feels all the quanta of energy but it applies no active force of its own upon them. The field (asymmetrically expanded space) simply presents an "inertial guiding path of least resistance" to each of the quanta and the quanta, individually, and by their own actions, force themselves down the guiding path toward the source for the field. This single quantum is considered as a constant in so far as its energy value (Planck's constant $h = 6.626 \times 10^{-34}$ joule-sec) is concerned. What does this say to us? Weight, relating to objects of mass is, at its essence, quantized. But note: the quantum's weight is a variable value which is ultimately determined by its location within the gravitational field. Note again that the gravitational field plays here but does not apply any active force.

But what value do we apply to this quantization. If we take a single particle of light far out in space beyond all gravitational fields its weight equals zero. But, if we position this same particle of light next to the "theoretical" black hole its weight might possibly be in the range of several tons. The three defining factors relating to an object's weight is 1. The amount of spatial expansion of the source's gravitational field, 2. The objects location within that gravitational field and finally 3. The number of the particles of light that object contains.

Mass and energy, according to theory, are the same thing in nature. Yet we see that the gravitational field does not recognize this theory. Or does it and we don't understand how? Using the equation $E = mc^2$, we see that when an object of mass, such as an electron, emits a single quantum h of energy it necessarily losses a small amount of mass. Some believe the mass of the electron is converted into energy but that is not the case. This quantum of energy, once emitted, has neither mass nor inertia but does keep its values of momentum and weight after ejection from the electron.

So what happened to the mass and inertia the photon absconded with? Going back; this quantum of energy, just before its emission, was traveling in an orbit within the electron, which is the action that generates the electron's gyre/inertia and mass. Within this work what we call electrons are in reality quarks. After emission, some of the gyre/mass and inertia is completely lost from the electron and the loosed quantum of energy is now in motion in a straight line somewhere thru space. Hence, the quantum of energy, no longer in an orbital configuration, can no longer drive the fabric of space surrounding the quarks in order to generate mass. Under this model, if no quarks are present you cannot have any value of mass. Understand that it takes orbital motion by kinetic energy within a quark to generate mass (angularly driven space) and this is why photons traveling in straight lines thru space have no value of mass.

This says thermodynamically, if we were to watch and measure the emissions from the electron over time, eventually the electron's gyre/inertia and mass would disappear and the final result would be many quanta of energy spreading out through the space around where the electron once existed. In addition to the laws of thermodynamics, all bodies of mass, are being constantly subjected to extraneous forces. All this together can be seen as entropy in action and it will eventually rob the mass from all matter in the universe which will leave only light.

Let us now "hypothetically" analyze the very last piece of mass, as it discontinues its orbital path characteristic and shows itself for what it really is, kinetic energy. Within this model or body of work, mass is simply a spatial wave side effect of kinetic energy orbiting within quarks. We know that the quantum value for energy is a constant h. If we assume that mass has no base value such as energy does, then we could, hypothetically, end up with not enough mass to create a full Planck h of energy as the electron reaches final decay. But Quantum Mechanics does not allow this result. Hence; under my model, every quantum of energy, in making one orbit within a quark, generates 1 quantum of mass in the space surrounding the quark. My model here shows very well what scientists already know; You can have a value of energy that does not contain mass but you cannot have a value of mass that does not contain energy. This says; light is a particle based entity while mass is a reactive spatial wave based entity.

And what else does this tell us? With reference to the equation $E = mc^2$, it tells us that kinetic energy determines the amount of mass any object contains. If an electron has X amount of mass and this mass is transformed into Y quantum amounts of energy, so that no partial amounts of mass X or energy Y are left, then, as I showed before, mass X must also have a base value h within it.

In reflecting back, the gravitational field cannot apply an active force but only offer an inertial guiding path of least resistance to the single particles of light that makes up the smallest constituent part (kinetic energy) making up the electron, separately from the electron as a whole, and this points to a conclusion that says, weight, mass, gyre and inertia must also be quantized. Note again; within this model quarks make up all mass matter and be it noted that electrons, and neutrinos, in that they have no nucleus, are in reality quarks. You can also say," any subatomic particle containing no nucleus must be classified as a quark".

To reiterate; weight within this model is a measurable and variable result of all the particles of light within bodies of mass forcing themselves to follow an inertial guiding path of least resistance thru space toward the source of the gravitational field they've encountered. A body of mass, without an external gravitational field surrounding it weighs zero.

I would ask of those who find fault with how I arrived at my results to overlook how I did it and look more at what I did. I simply changed Einstein's space-time continuum that curves in the presence of massive objects to an asymmetrical expansion configuration of the fabric of space, then inserted kinetic energy into an expansion generator equation, and finally rearranged the action of the forces involved. He constructed a physically curved space-time field while I constructed a physically expanded space. We both refused to classify gravity as a force. From there he postulated shortest possible paths while I postulated a spatial asymmetrically expanded inertial guiding path of least resistance.

In conclusion: no matter the process I used to gain my results relating to gravity, I believe I have discovered a great truth. Gravity cannot be labeled as an applied or active force, as has been assumed for hundreds of years. It can be said, and I believe, would be true; the gravitational field is simply an inert spatial framework that sits idly by, within the space surrounding bodies of mass, and the only vigor it shows is a passive, reflective, guiding, path of least resistance that it presents to any object that might become kinetically entangled within it.

CONCERNING THE THEORY OF THE EQUALITY OF INERTIAL AND GRAVITATIONAL MASS

In Einstein's book, "RELATIVITY- The Special and the General Theory", on page 66, he sets up an experiment with ropes and a large chest. He then puts a man in the chest and begins to accelerate the chest by means of the ropes attached externally. The man inside the chest doesn't know he is under acceleration and so, presumes he is sitting motionless within a gravitational field.

According to Einstein, this man can never figure out for sure, whether he is in an accelerated state or setting motionless within a gravitational field. This line of reasoning led Einstein to his Equivalency Law relating to accelerative forces and gravitational forces. But let us put this "Law" to a test.

First we exchange the chest with two space ships, call them N1 and N2. These ships have no windows so that anyone put inside the ship, will be unable to see out. Next we put an observer call him O1 within the ship N1. Then we put his identical twin brother in ship N2 as an observer call him O2. In addition, let us chose two points in space to test from. One is on the surface of the earth, call this frame of reference S1 and the other is out in space away from all gravitational fields call this frame of reference S2. The difference between the frames S1 and S2 is that in frame S1, the ship N1 is setting still on the surface of the earth and experiencing a gravitational force, call it G. In frame S2 the ship N2 is under an accelerative force call it A and equal in magnitude to the gravitational force G.

The observers O1 and O2 were unconscious when put into the ships so that when they wake up, they must then try and make an assessment as to which force, G or A, is acting upon them. We assume that they cannot hear the ship's engines of feel any vibrations. In order to let us better understand the process I will let it be known that ship N1 carrying observer O1 is setting motionless on earth at frame S1.

According to Einstein, observer O1 and O2 will not be able to determine which force G or A, is acting upon them. I disagree and will now show how observer O1 can tell which force G or A is acting upon him, while within the ship N1.

Since Einstein gives no time limit as to how long ship N2 can accelerate, I will have ship

N2 accelerate continually for the duration of the experiment. At the end of one month observer O1 on ship N1 has a real time photo taken of himself, and sends it off to his twin (observer O2) on ship N2 and observer O2 on ship N2 does the same. Observer O1 then knows that he is setting motionless on earth at frame S1. But now we must ask, "How can a picture of observer O1's twin brother let observer O1 decide which force G or A, is acting upon him.

I failed to mention that both twins were clean shaven when the experiment started. Once observer O1 seen that observer O2's beard was shorter than his, he immediately knew that his twin, observer O2 had been under the influence of acceleration.

On page 37 in the book, RELATIVITY - The Special and the General Theory – by Albert Einstein, he says, "As a consequence of its motion the clock goes more slowly than when at rest." This led me to conclude, "If a spaceship's clock runs slower when the ship is put into motion (accelerated) then so to would the biological clock, of the observer riding along with the ship, run slower."

Some might say that I cheated, relating to how I showed that observer O1 was under the influence of a gravitational field, but that is beside the point. I simply wanted to show that the effect of a gravitational field and the effect of acceleration were not equivalent when we factor in molecular action. I showed that even though we can't make the distinction, our atoms can. The important result is what I showed, not how I showed it.

In conclusion, I believe I achieved the result I sought and showed that the Equivalency Law for inertial mass and gravitational mass will not hold up to scrutiny. In conclusion I can now say that, "Acceleration and Gravity are <u>not</u> equivalent with respect to molecular action (quark reactions also)." In addition; put me on a space ship and let me guess whether I'm being accelerated or sitting still in a gravitational field. I will simply pull an electron from my pocket and if it's emitting light, above normal, I'm being accelerated, if not, I'm setting still within a gravitational field. Note: Science has been trying to figure out gravity for 400 years, it only took me 40.

CONCERNING GRAVITY AND THE BLACK HOLE

In 1916 Schwarzschild, with the help of Dr. Einstein and Laplace, constructed the first model of a black hole. According to the Encarta Encyclopedia, some astronomers believe that once an object collapses to within its Schwarzschild radius, it continues collapsing until it becomes a singularity, or a point with infinite density and a radius of zero.

Quoting from the book, MATTER, within the Life Science Library, by Ralph E. Lapp, page 179-180, "It has been known since the turn of the century that stars that have expended their thermonuclear energy turn into extremely dense bodies (called white dwarfs) as the gravitational forces of their atoms take over. For stars bigger than our sun, this aging process continues and they contract further into a core of even more densely packed neutrons. Neutron stars, weighing 10 billion tons per cubic inch, spin rapidly and emanate pulsating radiation ....

That this process of degeneration into ever-denser matter may, paradoxically, lead to the very annihilation of matter, was first suggested in 1939 by the atomic-bomb architect J. Robert Oppenheimer. Using Dr. Einstein's theory of relativity, Oppenheimer suggested that the gravitational collapse of a star continues until it squashes itself to death. Its matter disappears, entirely converted into pure energy. At the same time, its gravitational field becomes so strong that nothing, not even light, can escape it. Such a star would therefore be invisible - a black hole in space."

In the Microsoft Encarta encyclopedia, it says there was a theory put forth by John Wheeler, which Prof. Steven Hawking applied mathematics to, which says that matter entering a black hole loses its shape, chemical composition, and its distinction as matter or anti-matter, while still retaining its mass, angular momentum (spin) and electric charge. It took Heisenberg's uncertainty principle and borrowing virtual particles from empty space (nothing), to complete the mathematical model Hawking came up with.

The major problem with Black Hole theory is, a physical point with infinite density and a radius of zero (a singularity), would necessarily be infinitely solid and under such extreme gravitational contact within itself that no inner motions relating to mass or spin would be allowed. Put another way; there would not exist any space within the black hole to allow movement of any kind, which means all wave functions would disappear. Hence mass and inertia (internal gyre) would also disappear. Also, with a radius of zero, π would become unusable, hence with no radius, you can have no circumference. The question of how fast a Black Hole rotates or if it rotates at all becomes a moot point and ultimately unprovable. Mathematically, you have constructed a product having no factors with which to calculate. In reference to a singularity, Oppenheimer would experience the same problem.

I believe Oppenheimer was correct in saying that all matter within a black hole is converted into pure energy, but then we must adjust the theory to say, "Black holes are extremely dense objects within a very small volume of space. This approach allows us to at least stay within the laws of physics or reality, if you will.

Using Oppenheimer's theory gives us a picture of a very dense and very small black hole, full of only energy. Both Hawking and Oppenheimer conclude that a black hole will generate a gravitational field around itself having unimaginable force. The question becomes, "What is generating the extreme gravitational field?"

From the information above, we can come to only one assumption or conclusion. If an object is made up entirely by energy, then the energy itself, must generate the gravitational field surrounding the object. Dr. Einstein showed that the path light travels through the sun's gravitational field, is affected by the sun's gravitational field. Newton says that gravitation is a universal force, affecting all forms of matter and energy. Hence the light traveling by the sun also exerts a gravitational pull, though very small in strength, upon the sun. What does this tell us? It tells us that the generator for all gravitational fields is directly connected to a single quantum. When a single particle of light enters a gravitational field such as the sun possesses, the particle's kinetic momentum acts upon the field. Since the field (space) is asymmetrical in its density (expanded more the closer to the source you get) the particle of light is guided by its own application of momentum upon the field in a direction toward the source.

In conclusion, when an electron emits one particle/wave of radiation, it loses a small amount of energy, mass, weight, inertia and its gravitational potential. The particle/wave of radiation takes all these physical properties away with it when it leaves the electron. Hence, if this same particle/wave of radiation is absorbed by an atom, it will add energy, mass, weight, inertia and gravitational potential to this atom, at least until, it is reemitted by the atom.

As you can see, with reference to information above, energy itself generates the gravitational field and the more energy an object possesses, the greater becomes the gravitational field within and surrounding that object. The problem here becomes, if no form of matter or energy can escape a black hole then how, does a black hole generate such a powerful gravitational field around itself. And if the theoretical particle called the graviton exists then what allows it to escape and then be content to hang around the black hole? Many questions-few answers.

So now you ask, "How does all this apply to the idea of quantum gravity?" Let us use a proton as an example. We know, through the results of high energy physics, that the proton contains three very small but massive particles called quarks.

Newton seen three dimensions with time being separate and apart and configured his calculations as such. Dr. Einstein seen four dimensions, with time being the fourth dimension and configured his calculations as such. I see five dimensions with size being the fifth dimension and, with reference to analyzing quantum gravity, will configure my calculations as such. Note: QM theorists failed to add the size factor into their works regarding the subatomic world, hence they see electrons as having no radius (a point) and they also create equations that deny electrons a place in reality. In other words, an electron does not exist until you find it. Additionally, a principle of uncertainty is a mathematical guide post that must be followed on the subatomic highway. These are just a couple of the many car wrecks caused by not factoring in size when dealing with the quantum world. My model treats all this differently.

First we must set our sizes for the particles we will be working with. Follow here; The diameter of a hydrogen atom is approx. 1×10^{-8} m. The diameter of the proton is approx. 10^{-13} cm. The diameter of a quark is approx 10^{-15} cm. The diameter of an individual particle of space is undefined. And the diameter of a single particle (quantum) of light is approx. 10^{-24} cm.

Now, let us take an observer and shrink him until his height is 10^{-34} cm. Next we place this observer in the vicinity of a hydrogen atom (proton with one electron in orbit around it). We now have our observer calculate, by Newton's equation, the strength of the gravitational field he and the electron are experiencing.

He multiplies the mass of the proton by the mass of the electron and then multiplies this result by the gravitational constant G. Finally he divides this result by the square of the distance between the proton and electron. The force of attraction F between the proton and electron come out to be 1.6258^{-39} N. What does this tell us? It tells us that Newton's equation will not work at the quantum level, where the actual generator of gravity resides.

Recall before, when we showed that the gravitational force is a result of energy in motion within the elementary particles (quarks)? And where have we placed our observer? Alongside a proton which houses three quarks.

It is known that approx. 99% of the mass within a hydrogen atom resides within the nucleus (proton). Now since we know mass (particles of light moving in orbits) and gravity, are directly related, as we discussed before, this means our observer is experiencing a huge amount of gravitational force at his location. Just think about how massive the top quark is for its size and then realize the amount of energy it must contain.

In addition; if 99% of the mass resides within the nuclei, this means that 99% of the earth's gravitational field is generated by these same nuclei. The heaviest quarks are not gravitationally powerful enough to be a Black Hole, but they still generate a very substantial gravitational field around themselves, powerful enough, when combined with the other two quarks (proton), to trap electrons into orbits about them.

What our small observer finds is, the strength of the proton's gravitational field is immense or enormous, at this nuclear level or dimension, if you will. Remember, we are talking size here, of roughly 10^{-13} cm (Bohr territory).

The reason there is only one or possibly two electrons captured by the single proton, has to do with the standing wave created as the electron orbits the nucleus. Here I will refer you to Schrodinger's wave mechanics and to Pauli's exclusion principle and will not go into that further at this point.

What does this tell us? It tells us that the Strong Force, connected with the nucleus, can be defined as an effect directly connected to the quantum gravitational field surrounding the nucleus. As you can see, relativity, at the quantum level in nature, must not only factor in time (fourth dimension), but size (fifth dimension) as well.

To perform the necessary equation we would simply calculate how many quanta of energy are present within the area taken up by a quark and then, from what we discussed before, give each quantum of energy a measure of gravitational potential. When you combine the gravitational potential of all the energy within the quark and then factor in the size of the area taken up by the quark, you will find, you cannot keep light from escaping (Black Hole), but you can make the light run in circles to create a quark, like the Big Bang did.

It has been theorized that the strong force, within a proton, will change from an attractive force into a repulsive force at distances less than 2 femtometers. But as I see it, the mass (angular force waves propagating from the quarks) comes into play here. I will discuss these mass force waves in the next few paragraphs. Keep in mind that it is mandatory for me to explain all forces, and gravity also, as just a result (reaction) of energy in motion thru the fabric of space.

In a hydrogen atom, the quantum gravity generated by the quarks within the proton, causes the electron to take up an orbit close to the proton. The reason the electron does not fall into the proton is because, as we discussed before, the repulsive matter waves connected to the mass of both the quarks and the electron prevent the electron from coming too close to the proton. Hence the electron stays on the first step of the electron shells (ground state).

This is the mechanical configuration responsible for the configuration of a hydrogen atom. In addition, the more protons and neutrons a nucleus has, the greater becomes the nucleus's quantum gravity, hence the more electrons it can capture.

In the past it was assumed that an electron, in an orbit within an atom, could be considered to be under a constant acceleration. But mathematically, this showed that the atom would quickly experience complete decay. They were unaware that the electron was simply falling (inertial guiding path of least resistance) around the nucleus due to the effect of the nucleus's quantum gravity.

At this point it should be noted that Einstein wrote of his failure to unify the gravitational and electromagnetic fields into a one field configuration. I believe his failure came about mainly as a direct result of his denying the physical properties of space itself.

PART FIVE – ELECTROMAGNETISM

THE ELECTROMAGNETIC WAVE

At present electromagnetism is thought to be directly connected to light. This model agrees with that suggestion but only in a round-about way. My work is based upon the premise that all matter in the universe is essentially the result of energy in motion (quarks). So how do I generate electromagnetic fields under this configuration? A single quark, say an electron, put into motion will generate a frontal kinetic wave (deBroglie's pilot wave) within the fabric of space it is traveling thru. This is basically the generator of electromagnetism. So; light traveling in orbits generate a quark and a quark in motion generates a small electromagnetic field but when many electrons are put in motion such as they are when an electric current is applied to a copper wire a substantial electromagnetic field is produced. The resultant field is basically a result of space reacting to the electron's frontal kinetic wave in motion thru it. Even out in the galaxies, the fabric making up space being driven by interactions between bodies of mass in motion, is still the same generator of electromagnetism as is in the atomic world.

The electromagnetic wave is a resultant manifestation of the kinetic wave but shows itself, in strength, only by mutual electron action thru a conducting material. An individual electron, by moving through space and driving the space ahead of it gives itself what is known as DeBroglie's kinetic pilot wave. (See Figure 19). Note that the reactive standing wave following behind is not shown.

FIGURE 19.

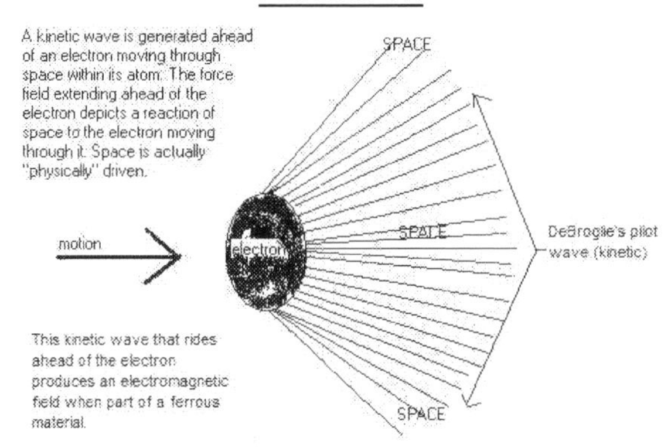

When groups of electrons are subjected to directional motion, such as in an electromagnet, the individual kinetic waves combine to create a strong driving of the space ahead of the electrons. This is the generator of the electromagnetic field and it can overcome the force of gravity for a short distance.

Oersted discovered, a magnetic needle is deflected at a right angle to current running through a wire. This showed the connection between magnetism and electricity. Oersted basically showed that an electric current could generate an electromagnetic field. Faraday later showed that a magnetic field in motion could also generate an electric current within a wire. Note that it took something in motion to generate both the field and the current. I will show that this "something in motion" is actually electrons in mutual alignment and mutual directional motion.

What Oersted discovered was electrons forced to move in unison and in the same direction through a wire (electricity), generates an electromagnetic field around the wire, causing the electrons within the compass needle to realign themselves perpendicularly, to the wire's electromagnetic field, which causes the compass needle to realign itself in space.

Faraday discovered that the electromagnetic field, which is generated by the motion of these electrons and is a physical extension of these electrons in motion, has the capability to force electrons into motion within a wire and create electricity. Note that the electron was yet to be discovered. I only wish that Faraday and Maxwell could imagine what was about to be discovered around the turn of the next century.

Electrons emit photons easily and since the generation of electromagnetic fields requires electrons to be in motion we will always have a side effect of light generation. Maxwell theorized after investigating Faraday's work and developing his field equations that light was actually the carrier for the electromagnetism. Einstein had not come on the scene yet with his theory showing photons as quanta of light particles. Maxwell saw light as a wave phenomenon. In this model, the spatial kinetic wave (deBroglie's pilot wave) associated with electrons in motion within wires is the generator of electromagnetism. And because space reacts at a speed equal to the speed of light, we can now understand how Maxwell discovered that the speed of propagation, for the electromagnetic field, equaled c. Note again though: electrons move at a speed always less than c but the fabric of space transmits its reaction to electrons in motion thru it, at a speed equal to but not related to the speed of light c.

ELECTROMAGNETISM

As is presently assumed, magnetism is produced by electric charge in motion. This "charge in motion" is the extent of the understanding science presently has for this phenomenon. This "charge" they speak of, is as I showed before, produced by individual electrons moving through space and "physically" driving the space ahead of them (deBroglie's kinetic pilot waves).

By wrapping a conductor of electricity (wire) around a ferrous rod and applying voltage to the wire we cause electrons to move through the wire. As these electrons travel through the wire they create an electromagnetic field about the rod. This field affects the atomic structure of the outermost atoms within the ferrous rod, which forces mutual alignment of the free electrons within the ferrous material (rod). By mutual alignment, I mean that the free electrons that the ferrous rod's atoms normally exchange between each other in all directions, is now exchanged in basically only one direction. This action aligns the directional motion of the electrons which drives the space ahead of them.

Normally these free electrons, within ferrous materials, are aligned haphazardly and no appreciable magnetic field is generated. This quickly changes as more and more electrons begin mutual alignment after being subjected to the force created by the electrons flowing through the wire, which by the way is also the source of electromagnetism. Electromagnetism, in its rawest form would be simply, an electron in motion. Also, a single electron in motion can be thought of as a mini-current with a mini-electromagnetic field being generated around it.

The spatial force, (kinetic waves) generated by the electrons moving through the wire, are the mechanical actions responsible for the mutual alignment of the electrons within the ferrous material. This action and the resulting motional alignment of the electrons cause the ferrous material to become an electromagnet, which then generates an electromagnetic field in the space immediately surrounding it. This electromagnetic field is the sum-total of all the biased kinetic wave action going on within the electromagnet and is itself a spatial kinetic wave, only much stronger than the individual electron waves. It can be thought of, as a combined wave of kinetic motion.

This kinetic wave drives space only at the negative end of the electromagnet and sets up what is called lines of flux, which is simply the physical configuration that space is being driven in. We must remember that Faraday and Maxwell constructed their ideas relating to electromagnetism with a physical space in mind. Heinrich Hertz later used a mathematical alternative to do away with the need for Maxwell's physical space. This I believe was a huge mistake and a precursor to the problems concerning infinities.

For every action there is an equal but opposite reaction and we see this at the positive end of the electromagnet. Because space, is being displaced at the negative end of the electromagnet, the space at the positive end is forced to move in to equalize this situation. And it moves into this diminished area of space at the positive end of the electromagnet, in approx. the same configuration as the space at the negative end is leaving. This creates flux lines at the positive end of the electromagnet that are similar to those at the negative end. Faraday showed by experimental means, that the flux lines of an electromagnet bend around to the positive end of the electromagnet in much the same configuration, as they hold on the negative end.

In the configuration I have described here, when another electromagnet is brought close to the one constructed here, if their poles are opposite, they are attracted to each other by way of the flux lines that space sets up around the electromagnet. The mechanical action that takes place in this situation is shown in Figure 20.

FIGURE 20.

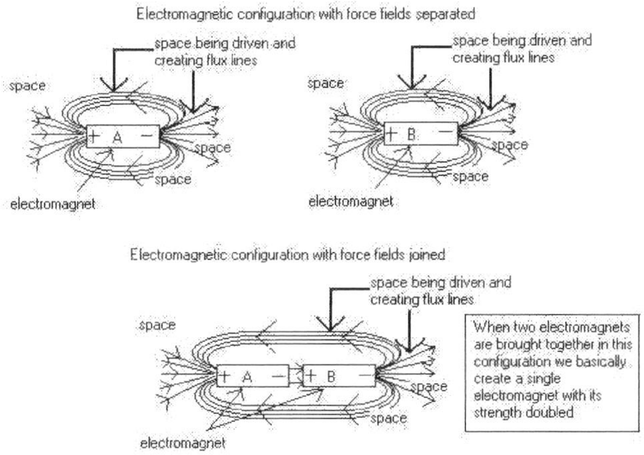

Electromagnetic configuration with force fields separated

space being driven and creating flux lines

space

+ A −

electromagnet

+ B −

Electromagnetic configuration with force fields joined

When two electromagnets are brought together in this configuration we basically create a single electromagnet with its strength doubled

Since the mechanical configuration of the electromagnetic field above (negative to positive) is a mirror image of the (positive to negative) configuration, I need not show it. But that is not true for showing the mechanical configurations that are created when bringing together electromagnets with like poles. It becomes necessary to show these like-pole configurations separately, due to the fact that the electromagnetic fields, space sets up around them, are not mirror images such as the opposite-pole configurations are.

When the negative ends of separated electromagnets are brought together close enough for their respective fields to start interacting, repulsion begins. This is due to space, being driven from the negative ends of both electromagnets as flux lines, colliding, and the reaction resulting from this, is repulsion between the electromagnets. The matter making up space, colliding, is the actual cause for this repulsion between the electromagnets. The flux lines colliding with other flux lines is an action and, by use of Newton's 3rd Law, this is felt back at the flux generating electrons. The electrons are in motion at some velocity < c but this velocity is not a direct factor in the projection speed of the resulting electromagnetic wave = c in the space around the electromagnet. So now we have the situation where the spatial force within the flux lines, upon collision with other flux lines of a like kind (force meets force), is reflected back towards the ferrous object from which it was generated. This is shown in Figure 21.

FIGURE 21.

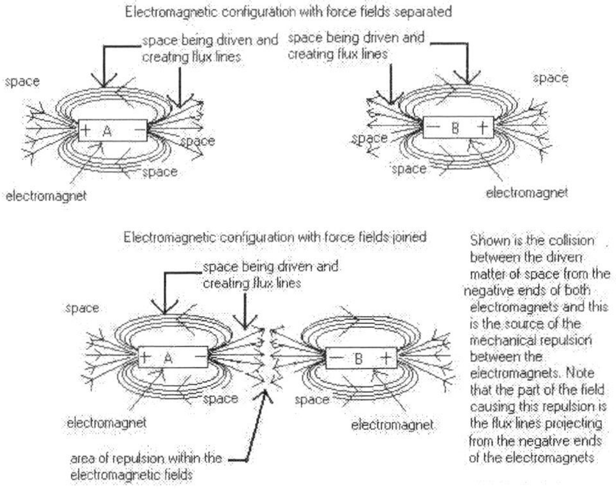

Electromagnetic configuration with force fields separated

space being driven and creating flux lines

space being driven and creating flux lines

space

space

+ A −

− B +

space

space

space

space

electromagnet

electromagnet

Electromagnetic configuration with force fields joined

space being driven and creating flux lines

space

+ A −

− B +

space

space

electromagnet

electromagnet

area of repulsion within the electromagnetic fields

Shown is the collision between the driven matter of space from the negative ends of both electromagnets and this is the source of the mechanical repulsion between the electromagnets. Note that the part of the field causing this repulsion is the flux lines projecting from the negative ends of the electromagnets

On the other hand, when the positive ends of separated electromagnets are brought together close enough, for their respective fields to start interacting, repulsion begins. This is due to the space, being driven from the negative ends, and then around to the positive ends of both electromagnets as flux lines, colliding, and the reaction resulting from this, is repulsion between the electromagnets. The fabric making up space, colliding, is the actual cause for this repulsion between the electromagnets. This is shown in Figure 22.

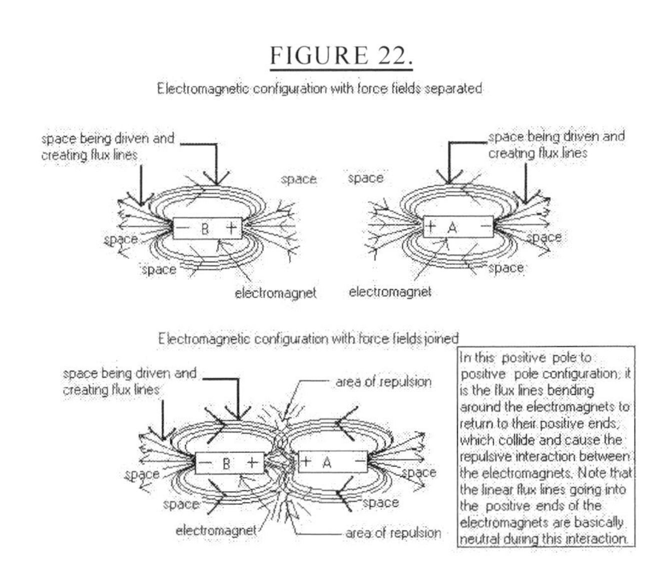

FIGURE 22.

The diagram in Figure 21. shows the negative repulsion occurring in the space directly between electromagnet A and electromagnet B. The diagram in Figure 22. shows the positive repulsion occurring along the outside of these same electromagnets. As a consequence of this, the repulsive strength of these electromagnets should be greater when in the negative to negative configuration as shown in Figure 21. I would think that this could be experimentally confirmed.

The scientific community presently believes that light is an electromagnetic wave. I have shown that in fact it's the other way around. The tiny kinetic waves riding ahead of their individual light particles orbiting within the quarks, when making up electrons, are the generators of the electromagnetic fields. Put simply, electromagnetism (result of space reacting to electron motion), is at its base, a force of orbital light within quarks.

ANDRE AMPERE'S EXPERIMENT

First, I will try to show that in the realm of electromagnetism, the theory of opposites exposes an anomaly. If we take a wire and cause an electrical current (electrons in directional motion) to run through it this will generate a biased electromagnetic field in the space surrounding the wire (right hand rule). If we then bring a compass near the wire the electromagnetic field surrounding the wire will cause the compass needle to align itself in an exacting manner (Oersted). This field surrounding the wire is considered to be one of the most basic forms of electromagnetism. Of course it is but what about the physical/mechanical nature of this field which surrounds the wire?

Andre Ampere found that when two metal wires were placed loosely, side by side, and with direct current flowing thru the wires in the same direction (negative and negative or positive and positive), they did not repulse each other but in fact moved closer to each other. I believe that if we configure the above under a physical/mechanical narrative we will arrive at only one conclusion; the electrons, while traveling along the wires, are driving the space itself ahead of them (electromagnetic field) on out away from the wires. Hence, this action forces the space surrounding the outside of the wires, to move in to replace the space that was driven out. This action is what causes the loose wires to come closer to each other. But be aware; excluding the original electron motion, what caused the wires to come together was not in fact, an attractive force that is particle based but rather a replenishing action of space itself by space itself. The wires came closer to each other simply because they were in the way of this replenishing action of space. Space itself is pushing the wires. In this case the Electromagnetic Fields are simply space itself being put into motion in response to electrons being put into motion. (See Figure23.)

FIGURE 23.

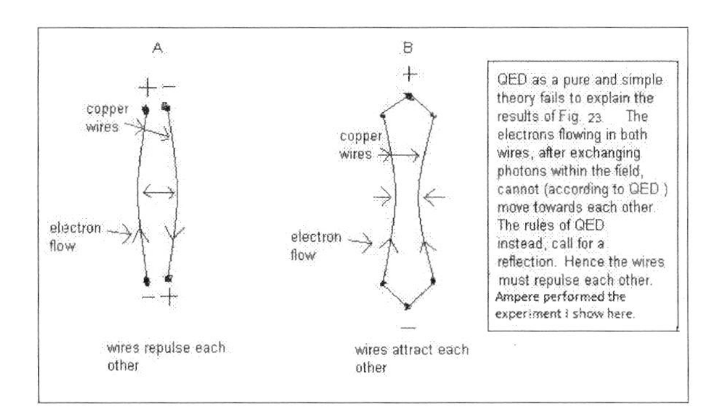

If we now reconfigure the setup so that one wire is negative and the other wire is positive the wires will be driven away from each other. This is because the electrons are driving the space surrounding one of the wires towards the space being driven by the electrons within the other wire. This colliding action causes the space between the wires to move outward which then pushes the wires outward. This produces the opposite result from what we showed just above.

To recap what's been said; when electrons within a conductive wire are put into mutual/ directional motion along the length of the wire, the space surrounding the wire is forced into motion which then generates what we call an electromagnetic field. Note also; Maxwell needed a medium for his electric and magnetic fields so he was forced to use space itself (then called the aether) to solve this problem. It was either Faraday or Maxwell, or both, who said, "The seat for electromagnetism resides within space itself". Maxwell went on to mathematically show that these fields propagate through the space around them at the speed of light. Hence he and others jumped to the conclusion that light might be an electromagnetic phenomenon.

Experimenters using iron filings show the flux lines going out from one pole of a magnet, bending around and then going back to the other pole. I have no doubt that when electrons are put in motion photons are ejected. My doubt is that these photons, which naturally move in straight lines, can also be the guiding force behind a bending electromagnetic field. In addition, this line of thought would necessarily lead us to conclude that the Big Bang (huge energy output event) was just one big electromagnetic happening. And I can't buy that.

PART SIX-IN CONCLUSION

THE REAL QUANTUM MECHANICAL REALITY

This concludes the initial presentation of my model. It must be given some weight by whoever reads it simply because of the numerous phenomena it mechanically explains. I would suggest that anyone reading this presentation now go back and review the foreword in order to gain an appreciation of just how many mysteries of science have been physically/ mechanically explained by this model.

I firmly believe that Michelson's aether experiment and its conclusion have set science back at least one hundred years. I would like to note that up until his death, Michelson still believed that the aether (physical space) existed. And in this belief, he was not alone. Newton, Lorentz, Planck, FitzGerald and many other great men of science also doubted the conclusion drawn from his experiment.

I personally refused to accept space as being a non-physical entity because all the accepted Laws of Physics cried out against it. What finally convinced me was the fact that light travels at c, independent of its source. This in itself says, the real cause for light's speed must rest with the particle of light and the environment around it. The idea that a radio wave with its small amount of energy, could maintain its speed for such a long period of time violated so many physical laws that I was forced to look elsewhere for this cause. This left only space, as a possible candidate for this cause, and it was only after I had exhausted all other, "energy" related approaches that I stumbled onto a possible solution. The Second Law of Thermodynamics defeated all my designs with the exception of "PRESSURE".

a. The pressure, space was subjected to, caused the Big Bang, and accelerated some of the particles of matter making up space, to c. This accelerated matter now became radiation, or what we call light.
b. Space pressure squeezes these light particles along at c while also restricting the speed of the light particles to c.
c. Space, in reaction to these particles of light in motion through it, generates a kinetic wave, in the space ahead of them. This is force in its rawest form, with the exception of pressure.
d. These kinetic waves, generated by the light particles moving under pressure, expand the space around them to create an "inertial guiding path of least resistance" or gravitational field.
e. These light particles, when orbiting within quarks, generate the force known as inertia or gyre.
f. These light particles and their associated kinetic waves, when orbiting within quarks, generate angular spatial matter waves, which can also be called the quantum force or mass if you will.
g. Matter waves, when part of an electron moving through its orbit around a nucleus, drives the space ahead of the electron to give the electron what is sometimes called deBroglie's pilot wave. Behind this pilot wave come reactive standing waves, (3rd law)
h. When a goodly amount of free electrons within a conducting material, are forced into directional alignment with each other, their kinetic space waves reinforce each other to cause a greater driving of space, and this action generates what is commonly known as an electromagnetic field.

This design could only be possible with a physical space. I don't need mention that this is also the only model that satisfies the perpetual requirement the universe must meet, in order to explain our own existence. We, and all matter around us are basically made from particles of light and their associated kinetic waves. We, ourselves are a Particle/Wave Duality.

"AND IT WAS SHOWN TO ME HOW LIGHT WAS BROUGHT FORTH FROM THE DARKNESS"

NUMBERS, DIMENSIONS AND TIME

On the planet earth within the milky-way galaxy there lives a species of animal known as homo-sapiens or humans, if you will. The humans have a somewhat advanced reasoning and data storage system encased within what is called a brain. The brain extends tendrils (central nervous system) throughout the body to control and regulate the various muscles, pumps, glands, sensors and filters needed to support and sustain it. Can we truly label robots as real??

With the above in mind, let us investigate what we call spatial dimensions. Many believe, without much thought I might add, that there are three spatial dimensions and in some cases more. In reality there exists only one true spatial dimension. The one true spatial dimension used to determine all other dimensions is length. Take a square object and start changing its orientation. You will find that height becomes depth, width becomes depth, and so on. This same process can be applied to a sphere. So; how do we determine the depth of an object? We measure the length of the depth. How do we determine the width of an object? We measure the length of the width. How do we determine the height of an object? We measure the length of the height.

In all three frames (height, depth and width) we have to use a length, with reference to a measuring device, in order to describe the size or shape of any object we encounter in nature around us. Man had to learn how to accomplish this feat. Human mental evolution progressed slowly when man was solely a hunter. Man was constantly on the move, following the herds with little time for anything else. Then man gained enough knowledge to become hunter-gathers. They now had extra time to ponder the universe around them. Their brains responded and grew in size and ability. I find it also possible that an outside influence might have been applied at some point in the past.

After some time (a later now) it became obvious that a way of describing things was needed so that all could understand things more fully. Guttural sounds, hand signs and ground or cave pictures alone would not suffice. This was the beginning of the scientific method (create/examine, test and verify). From this need man created, as a minimum, advanced languages, mathematics, geometry and time-keeping. Keep in mind that the ideas of time, numbers, laws and dimensions are all man-made measuring tools and are not automatically inbuilt characteristics with respect to nature itself.

I hope it is clear that even though theorists use such things as dimensions, numbers and time, these classical measurement systems fail to fully show us nature's underlying reality. Quantum Mechanics makes this very point. As for nature itself; there is only something under pressure continually generating motion within itself and the perceptible effect of this continuous action is what we have come to call "Now". What I have shown here is the dualism between human perception and reality. You must always be aware of this dualism as you use the scientific method in figuring things out. Believe nothing you hear and only a tenth of what you see!

THE CERTAINTY UNDERLYING THE UNCERTAINTY PRINCIPLE

In quantum mechanics there exists a principle supposing that it is impossible to simultaneously know both the location and momentum of a particle (i.e. electron) to an accuracy less than what the confines of Planck's constant allows. Experimentalists have made it very clear that this is indeed the case, at least for any classical probing technique.

But let us remind ourselves that just because we, as large classical biological observers, are unable to measure these sub-atomic values, that doesn't mean that these values don't exist in reality. An atom's electrons, at any one time, are all moving around the nucleus taking certain paths and with certain momentums. And they do exist as physical particles (quarks). This is a fact you can be certain of!

Remember that dimensions, numbers and time are classical man made measuring tools? In reality (the universe) has no dimensions. It simply is!! The universe as a whole has no center and also no edge, so how can up, down, length, width or any of this have a true point of reference apart from the universe itself. On the nuclear level under quantum mechanics, particles have been "theorized" to have no size (radius). I cannot imagine that such an impossibility might in fact exist! So how do we analyze elementary particle interactions? Hi-energy physicists rightly understand that force symmetry is the key to the quantum world. Let elementary forces become unbalanced and entropy immediately increases. Inspect and analyze the resulting disorder to figure out the previous order. The LHC at Cern was constructed for just this purpose.

As for time; it is directly tied to force. Let all the motion within the universe stop (dormant energy = no kinetic forces acting) and time becomes just a word with no meaning. In this situation all wave motion would also cease and all that would be left would be an infinite static space full of dormant particles of non-kinetic energy.

Let us examine a recent misuse of time. Feynman evidently thought that Tutankhamen, or someone, is still back in time riding his chariot and herding electrons from the past to the present and visa-versa. Tutankhamen, in reality, only resides in the here and now as a mummy in an Egyptian museum. And by applying QED to this situation we find that, "sure enough, Tutankhamen is now a mummy in an Egyptian museum". How uncanny!! Using QED, give me any answer and I'll find the question for it!

The point everyone seems to miss is that the future, past and present are all locked into what we call "now". The energy making up the universe is constantly transforming (under motion). We as humans, sense this transforming and have decided to measure it with a notched stick called "time". "Now" is what we call "present time". "Now" is a result of kinetic energy (light) in motion throughout the universe". The universe has internal momentum and it is a product of "light in motion" (kinetic energy). We and all other bodies of mass are a product of "kinetic energy interacting with physical space. This points to a conclusion that says; we are basically immortal in the fact that what makes up our composition (energy) is perpetual in nature. Sorry about your human future but glad for your energy's future.

By applying the principle of uncertainty, I think that there's a chance that somewhere out there someone will understand what I've shown here. But that remains uncertain.

CONCERNING THE ULTIMATE MIRACLE AND THE ULTIMATE LAW OF PHYSICS

First we must address the question; "Just what is a miracle"? The Encarta Dictionary defines a miracle as an event, occurrence or phenomenon that appears to be contrary to the laws of nature. If we analyze miracles as a whole, there exists only one true miracle and it relates to the fact that anything exists at all. Why not nothing? But yet here we are. Note: Without this first ultimate miracle all of existence would be impossible.

To go further, this miracle comes in two parts. The first part addresses the very existence of the universe. It's feasible that our universe could exist but be in a static state (no forces acting, entropy=zero). But we have found this is not the case. The second part of the ultimate miracle is the fact our universe is not static but contains and produces work (forces interacting, entropy>zero). Take note, without motion there can be no reason for human measurements of periodic change (time). Note also: If nothing existed (oxymoron) then the universe could not exist. There would be no energy, no mass or matter of any type, no forces, and, let me stress this "there would exist no space around and within everything".

At this juncture let us now analyze the connection between the ultimate miracle and the ultimate law of physics. Physics has but one law that deserves to be at the top of the scale. This law speaks for itself and can be considered as an Axiom (self evident truth). This law simply states that, "something cannot be obtained from nothing". And what does all this tell us? If we can't get something from nothing then logically:

1. The universe and the work it does, cannot have a beginning or end (It has to be both infinite and perpetual in nature).
2. The universe must be infinite in size, both macroscopically and microscopically which foregoes the need for the terms center and edge.
3. Space itself is simply dormant energy and when parts of it are released under pressure (Big Bang) it becomes kinetic energy (radiation). This outpouring of kinetic energy, under the effect of gravitational fields, then goes on to form various massive quarks from the top quark all the way down to the least massive quark known as the neutrino. Right after the Big Bang a small part of all the total radiation ejected was forced to join in orbits to create quark particles which then joined to create larger particles which then joined to make up

everything from neutrinos to galaxies. Note: Particles of light orbiting each other within quarks is what creates what we call gyre/inertia. This also explains the mysterious mass property that surrounds elementary particles. Note: Gyre "absolutely" points to internal motion of some type. How could they not see this! Also note Axiom #2: the universe cannot contain voids (areas filled up by nothing) within it.

BEYOND UNIFICATION

According to the most learned within the Scientific Community it is the Grand Aim of Science to define the nature of the Universe using the simplest and least descriptions possible. This can be done by asserting that the Universe is a conglomeration of substance excited by Pressure. This means that the movements the Universe makes is due to the pressure, the material, of which the Universe is made, exerts upon itself.

Let us take the Universe apart piece by piece and, while analyzing each piece, put it together again. First we do away with all life. Next goes all the visible and detectable parts such as particles, anti-particles, atoms, molecules, planets, moons, comets, rings, dust, gas, stars, galaxies, etc.. At this point all we have left is the space all this material once resided in and to make the picture complete we now do away with space itself. It is my intention to end up with "nothing", for this is the condition we must analyze before we can proceed.

The Universe under this condition would be non-physical or, to state it simply, would not exist. Under this set condition we have only one choice of mechanisms that could be used in manufacturing the Universe in which we now live. That mechanism would be the Universe actually jumping out of nothing or "Magic" if you will. Since I consider myself to be a physicist and not a magician I must discount the possibility of this "nothing" condition ever existing either in or before time.

So where do we start? From what we have just discussed it is obvious that we cannot start at the beginning. For according to the basic ideas of Physics, and especially the Laws of Thermodynamics, that "beginning" condition could not have existed when compared to the fact that if there was truly nothing before, then the Universe could not be here now. Since we cannot use a beginning condition for the study of the Universe we shall use a Local Prime Causal Condition. The choices for the make-up of this Prime Causal Condition are limited by our understanding of our local Big Bang itself. This line of reasoning necessarily leads to an assumption that a force, capable of causing an energizing action within the material body making up part of the universe itself, was somehow applied.

Taking all the known evidence collectively, we must necessarily admit that the output from the Big Bang was basically radiation in motion (kinetic energy). And of course motion bespeaks time. But motion of what? The answer can only be energy. And what may I ask, is the speed of this energy? The answer is 186,000 mps c. This by itself tells us that the speed of time in the early period right after the Big Bang occurred must have been equal to or close to c. This creates a conundrum where time is concerned.

We know that objects of mass travel at speeds less than c but this model being presented here shows kinetic energy as the basic generator of mass. In order to clarify this problem it must be understood that objects of mass (the quarks) that are made of light traveling in orbits within them at speed c, travel within space itself at speeds less than c. The universe basically has two configurations involving motion. In one configuration we see kinetic energy (photons) in linear motion everywhere within our universe. In the other configuration we see kinetic energy traveling in orbital motion which creates a spatial angular force I believe is mass (quarks). Put all this together and we end up with a situation involving two distinct time categories.

With reference to kinetic energy we have what I will label "active" particle time. With reference to objects of mass we have what I call "reactive" space time. Active time happens at a constant speed c while reactive time happens at various speeds less than c. But note that reactive time is a result of active time. I believe I've discussed time thoroughly enough, so lets move on. The Local Prime Causal Condition was without a doubt a result of the fabric of space being subjected to extreme pressure.

In giving the results of my investigation I feel it necessary to show why I could not have arrived at any other conclusion. As I have already stated, the idea of the Universe jumping out of nothing had to be rejected for reasons of logic as well as for requirements of physical laws. Note that science at present describe a point as an entity that exists even though it has no size (radius = zero). Note also that this hypothesis is based on assumptions in the quantum classical sense. But the mistake came when science applied this hypothesis to the quantum world. At 10^{-24}cm a point becomes far more than just a point.

I feel it necessary at this point to introduce the laws of thermodynamics and what they mean when applied to the Universe as a whole. The first law states that energy is conserved in all interactions, meaning that no machine can produce more energy than it consumes. When applied to the Universe this law does not allow a beginning for that would require the creation of energy from an unavailable material source (nothing). The main reason the scientific community refuses to speculate on what the condition of the Universe was before the Big Bang is that they assume that physical laws do not hold true, or apply, before that event. I intend to show that these laws governed even then. The First Law of Thermodynamics implies, if not proves, that the Universe could not have jumped out of nothing for that would have made it necessary for nothing to be made of energy. "NOTHING IS NOTHING" and therefore cannot have any physical value. The Law of Existence states: There cannot exist (take up room) a thing which has no value at all.

The Second Law of Thermodynamics states that energy is lost in any process and becomes unavailable energy. This law implies that the Universe, as a whole, should run down and eventually stop. Unless!!! The Universe has a built-in mechanism that could convert unavailable or used energy back to available or usable energy. This Second Law when combined (logically) with the First Law shows necessarily that the Universe could not have started and the movement within it must in fact be "PERPETUAL". Refer back to Part One if you have any questions.

One widely published theory holds that the Universe will continue to expand until it wears itself out and finally stops altogether. This theory we will call the Dissipation Theory. Another theory holds that gravity will eventually pull the Universe back into a singularity and "BANG" it starts all over again and so on in a cyclical fashion. This cyclical theory breaks down mainly due to the fact that calculations show gravity not having the strength or extension required to recall the radiation and light that precedes the expansion by an ever increasing distance. I have found that the inverse square law for gravity fails in the ability to recall all that original (Big Bang) light or radiation out there now beyond gravitational reach. Hence with each cycle the amount of material needed to reproduce the Big Bang would decrease until the point was reached where the system would stop. This fact combined with the requirements of the Laws of Thermodynamics show that a cyclical expanding and contracting Universe would have stopped at some point in the past and could not have started up again. The requirements of these laws also comes into conflict with the Dissipation theory mentioned above for the same reason. This is, however, not to say that the Dissipation theory is incorrect in so far as the wearing down of the expanding Universe is concerned.

Under further analysis I found that this problem went deeper. The Universe (as understood at present time), when subjected to the requirements of these laws, should have stopped a long time ago and should be stopped now. The problem was; the Second Law refused to let me start the Universe up again once it had stopped. Firmly lodged within the Second Law is the requirement that once the unavailable energy in any system equals the total energy, the system stops and unless additional energy is added to the system, in available form, the system will remain stopped. An analogy of this condition can be made using an automobile. This means that when all the gas turns into exhaust fumes the car stops and unless more gas is added, the car will remain stopped. In my case however, this means the Universe. If the Universe stopped there is no additional energy to restart it for the Universe already holds all the energy in existence. "I should note here that the definition of the Universe within this writing is: That, which is all of existence and contains all of existence." I tried to evade this (restart) problem for some time until I realized it was not a problem at all but an answer, to some of the most perplexing questions ever asked.

IN CLOSING

This theory is based upon the assumption that one cannot get something from nothing (see Axiom #1). It also holds as valid the laws of thermodynamics even when applied to the Universe itself. The following propositions are directly related to this theory:

1. The Earth is decreasing in mass/energy thus earth's gravitational potential is slowly decreasing.

2. The kinetic energy of which the bodies of mass within the universe are composed is slowly transforming, under pressure, back into dormant energy (space). This, my friends is entropy in action.

3. The Graviton, Magnetron and Gluon will never be found as they never existed.

4. What we call Gravity is not an active force at all. Gravity is a result of kinetic energy expanding the fabric of space it travels thru to produce inertial guiding paths of least resistance (Gravitational Fields).

5. Electromagnetic Fields are the direct result of the fabric of space reacting to the driving force of electrons in motion. The speed with which space transmits the shock applied to it, by electrons put into motion thru it, is equal to the speed of light but note that light is not the force carrier in this instance.

6. The fabric of space is the parent substance for all the visible and, so far, undetectable matter within the Universe.

7. The speed of a particle of light is regulated by the fabric of space which pressures it along while also regulating its speed.

8. The Universe is infinite in size and substance. Note that substance in this sense includes both a physical space and all the bodies of energy and mass within that space.

9. The Universe did not begin but always was, in some kinetic form or another.

10. When the Universe was less than 10^{-34} seconds old there was only the effect of a great pressure being released, as kinetic energy, from space, into space. Gravity was also spawned at this point.

11. The reason gravity is "assumed" to physically act upon all objects equally, no matter their weight or size, is because, in reality, the energy based gravitational field presents an inertial guiding path of least resistance to the individual particles of kinetic energy "within" matter and not, as is presently (2016 AD) thought, on the matter itself as a whole.

12. Singularities (points) are thought to have a radius of zero. That may be acceptable on a classical scale but on the quantum scale (10^{-24}cm), a point is very substantial in both its size and content.

I do not expect the Scientific Community to readily embrace this theory or even pay it much attention as there are already hundreds of theories out there dealing with much the same subject matter as mine. Therefore I find it necessary to copyright my works as a sort of time capsule against the day when the Universe is better understood and some young physics or astronomy student stumbles onto my writings. I fully expect the day to come when Science will once again demand empirical facts for every theory or hypothesis submitted. Then shall they take up Occam's Razor to peel back the absurdities and let the true light shine through in its own simple brilliance. I attempted to combine the workings of the Universe under one force (pressure) and believe that I have succeeded. All my findings are based on the belief that you cannot get something from nothing. This one belief was the gauge against which I tested all that appears within my theory. I will now give a proposition of a different nature which has to do with the regulated speed of the movements within the Universe.

Everything, with the exception of mass objects, move at the same rate within the Universe. We, as objects of mass travel within (not thru) space at speeds less than the speed of light c. Einstein set up the framework for this proposition when he proposed that clocks will run slower as they approach the speed of light. This would lead one to conclude, under the present misconception on empty space, that particles must be able to read a speedometer in order to allow themselves to increase in mass and slow down their internal workings. The idea that the governing factor for this phenomenon is an isolated internal device within all particles is absurd. As an example for this, consider the fact that light slows down when traveling through a medium such as the Earth's atmosphere. Once it is out of the atmosphere though, it speed rises back to 186k mps, "supposedly" instantaneously. What tells the photon to slow down? We know this to be a ridicules question for we know that it is the medium or atmosphere that is causing the slow down. But, on the other hand, what tells the photon to speed up once it has traveled through the atmosphere and back into empty space. Have we now given Maxwell's Demon a new responsibility? I think we can safely assume that no little creature of any sort rides inside the photon in order to regulate its speed. In reality the light never slows down but in fact, it just takes more time for the light to travel the non linear path it must take in passing thru the medium.

To go further; why does light travel so slow in empty space which, supposedly, can offer no resistance? Why not a million miles per second or a trillion? The answer must necessarily end up to be: The speed of all particles is regulated by the forces or medium surrounding it. In the case of light in "presumed" empty space, it has to be space itself which both accelerates and regulates the light's speed. This idea, when scrutinized closely, brings much doubt to bear on the validity of light regulating its own speed. To the students reading this theory in 3000 AD: by now you must know that the speed of light is not a true constant, as is presently thought, and can be altered by where, in space, you are making the measurements. Back in the 20th and 21st centuries I felt like a child crying at the darkness; I can't see you but I know your there!"

The Michelson Experiment, which tried to detect space itself, becomes contaminated once it becomes understood that the tool used (light) is actually regulated by what he was trying to detect. This reminds me of the Galileo event when he tried to measure light using shutter-lanterns and his heart rate. What is equally hard to believe is the fact that any study of light itself, once analyzed, would point to only two alternatives as choices for regulating its speed. Those choices are either magic or a medium. I believe that space itself is that medium.

In closing; I would like to believe that I have presented my work in an understandable manner. Mathematics have been held to a minimum for a good reason. I have tried to show the Universe as an eternally motivated object. Logically, I believe that the very act of my being able to exist and to produce the energy to write this paper should be proof enough. I also believe that any student of science doing a study of the Universe and adhering to the Three Laws of Thermodynamics will eventually come to the same conclusion as I have. How do you end something that never started or stop the movement it makes?????????

In the world of physics today there are many interactions taking place that we find that we are unable to explain. These are normally called phenomenon. Gravity and the speed of light are just two of the many. We also find that the mathematical road that leads to the quantum world has serious deficiencies in that, fact is often stretched to fit theory and in some cases the road is missing altogether. By this I mean that if experiment does not verify a particular theory often times the experiment is concluded to be at fault and not the theory. This is particularly so if the theory in question happens to be based upon one of the "Sacred Cows" of science. As an example one would only have to look as far as renormalizing equations or dealing with infinities. Although both showed the deficiencies of many theories, it was not the theories that were changed but the way we deal with renormalization and infinities. Mathematical card tricks were used in many instances to circumvent these problems. As a result physics today finds itself anchored by a chain with many missing or weak links.

CONCERNING PHYSICS WHEN APPLIED TO LIFE AND DEATH

Using a human biological frame of reference, a body can be considered to be alive if both metabolism and reproduction are active within that body. NASA, I was told, uses a guideline for identifying life which says, "if an object can move or function of its own resolve in an orderly fashion it must be considered a strong candidate for containing life".

Expanding upon this line of reasoning, the event known as death, with reference to human understanding, occurs when all vital functions or interactions terminate within an organism or cell. But duly note; all the above is based upon mental human concepts and observations. Humans believe in what they see as life and death. I will now show that, by the laws of physics and common sense logic, the idea of life and death is the result of a false con-struct which was developed solely from ignorance.

No doubt this false con-struct came about as early humans witnessed what they thought to be life and death happening all around them. In truth, their beliefs in life and death were valid but only with reference to themselves. With reference being existence as a whole, plant and animal life can simply be categorized as but, one out of many, energy transformations that has happened over time for billions of years. Always keep in mind however, existence itself has no beginning or end and it full well knows that.

What's being pointed out here is the fact that man is oblivious of the true universal concept of life and death. Humans see and experience their life and death while the universe as a whole does not. The universe sees only transformations involving interactions between space and the kinetic energy in motion within this same space. With reference to the conservation of energy law we can say, "Humans are conceived, then live and die, but the universe itself experiences no gain or loss of energy during this event". So now comes the question; "is nothing lost then, when one dies?" Let us examine this question.

The universe as a whole, I am sure, is aware of its own existence and its ability to manipulate the goings-on within itself. On the other hand, some of the goings-on can be manipulated by entities within. What I speak of here is the life forms that exist within the universe. These life forms, I believe, can act of their own will without guidance from outside entities. This says simply, we have free will, separate and not connected to the universe as a whole. Put bluntly; life itself is a "special force" of nature.

Again; humans see what they think is life and death and believe as much. But an insensible and sterile universe sees not life nor death but only active timely events resulting from energetic transformations. Einstein's relativity, under the guise of a mental-biological human frame of reference, might declare that both views are equally true. But I disagree and put it bluntly to all persons of ample knowledge, "With reference to the conservation of energy law, no part of the universe as a whole can truly die, just transform". Using just the laws of physics this says, the universe as an insensible sterile entity does not recognize life and death at all but only the transformations the energy is experiencing. The universe itself sees you not by your name or "temporary" form but by the radiation signature you trace out in the space surrounding you. This reality might be the very reason we humans feel the need for a GOD. I know it is true for me. I believe the universe is infinite and I also believe it possesses a consciousness along with an awareness of self. I believe the universe itself is alive and possesses self-awareness. But that is only my personal belief for I also believe everyone must seek their own personal deity or GOD if you will.

All bodies of matter (mass), will eventually be found to be finite in time itself, but energy, on the other hand, is perpetually infinite and this is the key to accepting our fates if you will. We humans along with all other objects of mass within the universe are truly immortal, just not in the sense we would prefer. Note: you can have energy without mass being present, but you cannot have mass without energy being present. High energy collisions and anti-matter interactions show this result to a high degree of certainty. Let me state, "you might believe John is dead and gone, when in fact, his energy, but not John, will go on and on".

In conclusion: humans are basically wave functions within space which are a result of the actions of kinetic energy interacting with space itself. Consciousness and thinking can be reduced mostly to electrochemical interactions within the brain. The DNA contains our controlling software. I have no doubt that the earth's plant and animal DNA codes were written (created) by entities somewhere out there within our local group of galaxies. Life itself is a force though it is special in that it usually has the ability to apply itself as it wishes.

But do not for a minute think that what I've presented here is a barrier to my belief in God. My hypothesis here plainly says, "We humans will eventually lose our ability to self-motivate but that does not equate to a true universal death". The energy within us is perpetual. By stepping outside of our human perspective there is much more to be discovered. Again; Look for that which dwells beyond the obvious and still "Question Everything"!!!

LAST NOTE

1. **I could not accept that existence itself had a beginning!**

2. **I could not accept that space itself was a void!**

3. I could not accept that light goes from zero to c instantaneously!

4. After that I realized everything was fair game!

TO PUT IT ALL IN PERSPECTIVE: THE PERPETUAL AND INFINITE UNIVERSE IS THE ACTION WHILE WE AND ALL OTHER FORMS OF MATTER ARE SIMPLY REACTIONS.

Printed in the United States
By Bookmasters